Biomechanics of the Gravid Human Uterus

Roustem N. Miftahof · Hong Gil Nam

Biomechanics of the Gravid Human Uterus

Springer

Professor Dr. Roustem N. Miftahof
Arabian Gulf University
College of Medicine and Medical Sciences
Department of Physiology
26671 Manama
Kingdom of Bahrain
roustemm@agu.edu.bh

Professor Dr. Hong Gil Nam
Pohang University of Science and
Technology
Plant Systems Bio-Dynamics Laboratory
San 31, Hyoja Dong, Nam Gu
Pohang, Gyungbuk, 790-784
Republic of South Korea
nam@postech.ac.kr

ISBN 978-3-642-21472-1   e-ISBN 978-3-642-21473-8
DOI 10.1007/978-3-642-21473-8
Springer Heidelberg Dordrecht London New York

Library of Congress Control Number: 2011933468

Cover design: deblik Berlin, Germany

Printed on acid-free paper

Springer is part of Springer Science+Business Media (www.springer.com)

Preface

Thus, the task is, not to see what no one has yet seen, but to think what nobody has yet thought, about that which everybody sees.

E. Schrödinger

The complexity of human uterine function and regulation is one of the great wonders of nature and it represents a daunting challenge to unravel. Our current understanding of uterine functions is based mainly on information obtained from in vivo and in vitro experiments as well as subsequently derived concepts presented in a form of "theoretical" models. Most of the research on human myometrium has been aimed at the elucidation of biochemical pathways for contractility. Ethical issues, the lack of standardized methodologies, and a big scatter of experimental data make the performance of such studies and the interpretation of the results difficult. Although experiments are supplying and will continue to offer valuable data at different structural levels on myometrial activity, they fail, however, to provide a holistic view on how the uterus works. Therefore, it is time to bring together previously unconnected pools of knowledge as an integrated system.

The use of mathematical ideas, models, and techniques is rapidly growing and is gaining prominence through the biosciences. The field of computational systems biology emerged from the need to integrate multicomponent biological systems and establish missing functional links among them. Applied mathematicians and bioengineers working alongside bioscientists provide a quantitative description to intricate processes at the subcellular, cellular, and tissue levels and integrate them in "viable" models. Such models uphold gnostic properties, offer an invaluable insight into hidden and/or experimentally inaccessible mechanisms of organ function, and allow the researcher to capture the essence of dynamic interactions within it.

At the time of writing of this book, publications dedicated to biomechanical modeling of the gravid uterus were sparse. We aimed to fill the gap and to give an example of application of mechanics of solids and the theory of soft shells, in particular, to study mechanics of the pregnant uterus. We attempted to knit the precise science of mathematics and the descriptive science of biology and medicine into a coherent structure that represents accurately the way they should interact.

The breadth of the field, which is constantly expanding, forced a restricted choice of topics. Thus, we focused mainly on electromechanical wave processes, their origin, dynamics, and neuroendocrine and pharmacological modulations. Naturally, no pretense is made that every facet of the subject has been covered, but we have tried to give a consistent treatment of each.

The book serves three functions: first, it introduces general principles of construction of a multiscale model of the human womb by integrating existing experimental data at different structural levels; second, the book provides, through numerical simulations, the understanding of intricacies of physiological processes, and third, it puts current knowledge into perspective. We assume that the reader has an adequate knowledge of human physiology, biochemistry, pharmacology, mathematics, and mechanics of solids such as might be gained from general courses at university. Therefore, we believe that the book should serve as a text for courses in computational systems biology for advanced undergraduate and/or first year graduate students, as well as researchers and instructors of applied mathematics, biomedical engineering, computational biology, and medical doctors.

A brief overview of the anatomy, physiology, and mechanics of the uterus which is sufficient to follow the arguments is given in Chap. 1. One should not expect, though, to find a complete biomedical survey on the subject. The interested reader is advised to consult special literature for that. Emphasis is rather given to adaptation and revision of known experimental facts to formulate constructive hypotheses, justification, and rational selection of models and their integration from the molecular up to the organ level.

The current trend in mathematical modeling of myoelectrical and mechanical activity in myometrium and the uterus is outlined in Chap. 2. Existing, mainly null-dimensional, models focus on the analysis of cellular mechanisms and the identification of regulatory factors responsible for bursting and contractility. A few models have been developed to simulate the organ as a thin elastic membrane. Under general assumptions of geometrical and physical nonlinearity, attempts have been made to estimate the dynamics of labor. However, the existing models serve specific needs, i.e., they are designed to explain certain in vitro experiments or to justify quantitatively particular clinical measurements, rather than offering a solid framework for the development of a general theory of uterine function.

Throughout the book we adopted a phenomenological (deterministic) approach to describe biological phenomena under investigation. The concept of a functional unit that was successfully applied to study biomechanics of other organs (Miftahof et al. 2009; Miftahof and Nam 2010) is employed to model the gravid uterus. A one-dimensional model of the myometrial fasciculus based on morphostructural, electrophysiological, and biomechanical principles of function is developed and analyzed in Chap. 3. A special attention is given to analysis of the generation and propagation of the electromechanical wave of depolarization along the myofiber under normal physiological and chemically altered conditions.

Basic concepts of the theory of surfaces which are essential to a subsequent understanding of the mathematical model of the uterus are discussed in Chap. 4. We have retained an emphasis on correctness and depth of conceptual arguments

without recourse to advanced mathematics. Thus, the reader should find it quite
easy to apprehend. Throughout the book we used tensor notation, which is the most
"economical and concise" way of bringing the discussion to the point.

The following Chap. 5 is fundamental for the model development. It introduces
a class of thin soft shells that possess zero flexural rigidity, do not withstand
compression forces, experience finite deformations, and their stress–strain states
are fully described by in-plane membrane forces per unit length. Since three-
dimensional equations for a shell are generally very complicated, the aim of the
shell theory is to reduce these equations to two-dimensional form. This reduction is
achieved by formal integration of the equations of equilibrium of a three-dimensional solid over the thickness of the shell and a successful application of the second
Kirchhoff–Love hypothesis. As a result, the equations of motion of soft shells are
derived in general curvilinear coordinates.

Chapter 6 is dedicated to constitutive relations for the myometrium. It is treated
as a chemically reactive mechanical continuum. By examining the phenomenological bases for the equation of state together with elements of thermodynamics,
a connection between macroscopic and microscopic descriptions in the myometrium is established. The fundamental concepts and their ramifications are presented in ways that offer both their significance and biological validity. At the end
of the chapter, the method of calculation of membrane forces in the principal
directions is given.

In Chap. 7, we developed a theoretical framework for the analysis of integrated
physiological phenomena in the pregnant uterus conveyed by multiple neurotransmitters and modulators. The emphasis is given to modeling of receptor polymodality, coexistence, and co-activation of different families of G-protein second
messenger systems and intracellular signaling pathways. We concerned ourselves
with the spectrum of responses exerted by acetylcholine, adrenaline, oxytocin,
prostaglandins, and progesterone acting alone or conjointly.

An increased knowledge of the molecular and physiological mechanisms implicated in the genesis and pathogenesis of human labor aids the identification of new
targets sites through which to modulate myometrial contractility. Preterm labor
carries a wide range of problems, which vary from acute – respiratory distress,
intraventricular hemorrhage – to long-term handicap–cerebral palsy, visual, and
hearing impairment. However, the development of novel therapeutics to treat the
preterm birth is being hindered by our incomplete understanding of signaling
pathways that regulate the process. In Chap. 8, we analyzed pharmacokinetics of
major classes of drugs used in clinical practice to manage preterm labor and
associated conditions. Mathematical models of competitive antagonists, allosteric
interactions, and allosteric modulation of competitive agonist/antagonist actions
were proposed.

In Chap. 9, we give a mathematical formulation of the problem of electromechanical activity and neurohormonal regulation in the pregnant human uterus
and present results of some numerical simulations. Accurate quantitative evaluation
and comparison to in vivo and in vitro experiments were not possible. Therefore,
we focused on qualitative analysis of processes in the organ during different stages

of labor. We demonstrated that during the first stage the symmetry and synchronicity in electromechanical activity is the most beneficial. However, a delay in excitation and the spiral wave formation in the circular myometrial syncytium are essential in the second stage for rotation of the engaging part of the fetus. The third stage of labor could not be assessed with the proposed model because of the significant changes in constructive characteristics of the postpartum uterus. We did not intend to analyze in detail pathology of labor; however, we thought that it would be beneficial to simulate a few common conditions, e.g., constriction ring, hypo- and hypertonic inertia, and uterine dystocia.

The last Chap. 10 briefly outlines ontologies of modeling and discusses applications, pitfalls, and problems related to modeling and computer simulation of the human pregnant uterus and pelvic floor structures.

A collection of exercises are included at the end of each chapter to enable the reader to expand their knowledge in the subject and to acquire facility in working out problems.

We hope that the book will convey the spirit of challenge in application of mathematical methods to biomedicine and bring to light new concepts, targets, and expectations by illustrating the breadth and complexity of the field, highlighting problems and achievements already made, and by demonstrating the immensity of the future task in unraveling mysteries of the human uterus.

Finally, we would like to thank our friend and colleague, Dr. W. Morrison, for his invaluable assistance in reviewing the manuscript. We owe a debt of gratitude to our families and friends for their forbearance and encouragement. We also extend our gratitude to Dr. Andrea Schlitzberger and the staff at Springer, Heidelberg, Germany, who have supported the project from the beginning and have helped in bringing this book to fruition.

R.N. Miftahof
H.G. Nam

Contents

Notations

$\overset{0}{S}, S, \overset{*}{S}$ — Cut, undeformed, and deformed (*) middle surface of a shell

h — Thickness of a shell

x_1, x_2, x_3 — Rectangular coordinates

r, φ, z — Cylindrical coordinates

$\{\bar{i}_1, \bar{i}_2, \bar{i}_3\}$ — Orthonormal base of $\{x_1, x_2, x_3\}$

$\{\bar{k}_1, \bar{k}_2, \bar{k}_3\}$ — Orthonormal base of $\{r, \varphi, z\}$

$(\alpha_1, \alpha_2), (\overset{*}{\alpha}_1, \overset{*}{\alpha}_2)Z$ — Curvilinear coordinates of the undeformed and deformed shell

$\bar{m}, \overset{*}{\bar{m}}$ — Vectors normal to S and $\overset{*}{S}$

$\bar{r}, \bar{\rho}$ — Position vectors

$\left.\begin{array}{c}\{\bar{r}_1, \bar{r}_2, \bar{m}\} \\ \{\bar{r}^1, \bar{r}^2, \bar{m}\}\end{array}\right\}$ — Covariant and contravariant base at point $M \in S$

$\chi, \overset{*}{\chi}$ — Angles between coordinate lines defined on S, $\overset{*}{S}$

γ — Shear angle

A_i — Lamé coefficients on S

a_{ik} — Components of the metric tensor

$a, \overset{*}{a}$ — Determinants of the metric tensor

b_{ik} — Components of the second fundamental form

$ds, d\overset{*}{s}$ — Lengths of line elements on S, $\overset{*}{S}$

δs_Δ — Surface area of a differential element of S

$\Gamma_{ik,j}, \Gamma^j_{ik}$ — Christoffel symbols of the first and second kind

A^j_{ik} — Deviator of the Christoffel symbols

R^j_{ik} — Riemann–Christoffel tensor

$\bar{\upsilon}(\alpha_1, \alpha_2)$ — Displacement vector

ε_{ik} — Components of the tensor of planar deformation through points $M \in S$

$\tilde{\varepsilon}_{ik}, \overset{*}{\tilde{\varepsilon}}_{ik}$	Physical components of the tensor deformation in undeformed and deformed configurations of a shell
$\varepsilon_1, \varepsilon_2$	Principal physical components of the tensor of deformation
$\varsigma_{ij}, \Delta_{ij}$	Elastic and viscous parts of deformation
$\lambda_i, \lambda_{c,l}$	Stretch ratios (subscripts c and l are referred to the circular and longitudinal directions of a bioshell)
Λ_1, Λ_2	Principal stretch ratios
$I_1^{(E)}, I_2^{(E)}$	Invariants of the tensor of deformation
$\left. \begin{array}{l} e_{nn}, e_{n\tau}, e_{n\tau} \\ e_{\tau n}, \omega_n, \omega_\tau \end{array} \right\}$	Rotation parameters
$1/R_{1,2}$	Principal curvatures of S and $\overset{*}{S}$
K	The Gaussian curvature of S and $\overset{*}{S}$
$e_{\alpha 1}, e_{\alpha 2}$	Elongations in direction α_1, α_2 respectively
p	Pressure
$\bar{p}_i$	Stress vectors
$\bar{R}_i$	Resultant of force vectors
$\bar{P}_{(+)}, \bar{P}_{(-)}$	External forces applied over the free surface area of a shell
$\bar{F}$	Vector of mass forces per unit volume of the deformed element of a shell
$\overset{*}{T}_{ii}, \overset{*}{T}_{ik}, \overset{*}{N}_i$	Normal, shear, and lateral forces per unit length
$T^t, T_{c,l}$	Total force per unit length of the fasciculus and myometrium, respectively
$T_{c,l}^p, T_{c,l}^a$	Passive and active components of the total forces per unit length
T_1^r, T_2^r	Forces per unit length of reinforced fibers
T_1, T_2	Principal stresses
$I_1^{(T)}, I_2^{(T)}$	Invariants of the stress tensor
σ_{ij}	Stresses in a shell
σ_{ij}^α	Stresses in the α phase of a biomaterial
$c_1, ..., c_{14}$	Material constants
d_m, d_f	Diameter of smooth muscle fiber and nerve terminal, respectively
L, L^s, L_0^s	Length of bioshell/muscle fiber, axon, and nerve terminal, respectively
k_v	Viscosity
ρ	Density of undeformed and deformed material of a shell
ρ_ζ^α	Partial density of the ζth substrate in the α phase of a biomaterial
m_ζ^α	Mass of the ζth substrate in the α phase of a biomaterial
$\upsilon, \upsilon^\alpha$	Total and elementary volumes of a biomaterial

c_ζ^α	Mass concentration of the ζth substrate in the α phase of a biomaterial
η	Porosity of the phase α
Q_ζ^e, Q_ζ	Influxes of the ζth substrate into the phase α, external sources, and exchange flux between phases
$v_{\zeta j}$	Stoichiometric coefficient in the jth chemical reaction
$U^{(\alpha)}$	Free energy
$s^{(\alpha)}, S_\zeta^1$	Entropy of the α phase and partial entropy of the entire biomaterial
T	Temperature
μ_ζ^α	Chemical potential of the ζ th substrate in the α phase of a biomaterial
$\overline{q}$	Heat flux vector
$\mathbf{R}$	Dissipative function
Λ_j	Affinity constant of the jth chemical reaction
$\overline{J_i}, \overline{J_o}$	Intra- (i) and extracellular (o) ion currents
I_{m1}, I_{m2}	Transmembrane ion currents
$I_{ext(i)}$	External membrane current
I_{ion}	Total ion current
$\left. \begin{array}{c} \vec{I}^s_{Ca}, \tilde{I}^f_{Ca}, \tilde{I}_{Ca-K}, \tilde{I}_K \\ \tilde{I}_{Cl}, I_{Ca}, I_{Ca-K}, I_{Na} \\ I_K, I_{Cl} \end{array} \right\}$	Ion currents in myometrium
Ψ_i, Ψ_o	Electrical potentials
V_m, V_p, V^s, V^f	Transmembrane potentials
$V_{c,l}, V_{c,l}^s$	Membrane potentials in circular and longitudinal myometria
V_p^s	Threshold potential
$\tilde{V}_{Ca}, \tilde{V}_K, \tilde{V}_{Cl}$	Reversal potentials for Ca^{2+}, K^+, and Cl^- currents in myometrium
V_i	Membrane potential of the pacemaker zone
$\left. \begin{array}{c} V_{Ca}, V_{Ca-K}, V_{Na} \\ V_K, V_{Cl} \end{array} \right\}$	Reversal membrane potentials for pacemaker zone cells
$V_{syn}, V_{syn,0}$	Actual and resting synaptic membrane potentials
$V_{Na}^f, V_K^f, V_{Cl}^f$	Resting membrane potentials for ion channels at the synapse
C_m, C_m^p, C_p, C_a^f	Membrane capacitances of myometrium, fasciculus, synapse, and axon, respectively
$R_{i(0)}^{ms}$	Membrane resistance
R_s, R_v, R_a^f	Specific membrane resistances of a fasciculus, synapse, and axon, respectively
$\hat{g}_{ij}, \hat{g}_{oj}$	Intra- (i) and extracellular (o) conductivities
$\hat{g}_{i(o)}^*$	Maximal intra- (i) and extracellular (o) conductivities

$$\left.\begin{array}{l} g_{Ca(i)}, g_{Ca-K(i)}, g_{Na(i)} \\ g_{K(i)}, g_{Cl(i)}, \tilde{g}_{Ca}^f, \tilde{g}_{Ca}^s \\ \tilde{g}_K, \tilde{g}_{Ca-K}, \tilde{g}_{Cl} \end{array}\right\}$$ Maximal conductances of respective ion channels

$g_{Na}^f, g_K^f, g_{Cl}^f$ Maximal conductances of ion channels at the synapse

$$\left.\begin{array}{l} \tilde{m}, \tilde{h}, \tilde{n}, \tilde{x}_{Ca} \\ h_{Na}, n_K, z_{Ca}, \rho_\infty \end{array}\right\}$$ Dynamic variables of ion currents in myometrium

m_f, n_f, h_f Dynamic variables of ion channels at the synapse

$\tilde{\alpha}_y, \tilde{\beta}_y$ Activation and deactivation parameters of ion channels

$Z_{kl}^{(*)}$ "Biofactor"

$[Ca_i^{2+}]$ Intracellular concentration of free Ca^{2+} ions

ϑ_{Ca} Parameter of calcium inhibition

$\hat{\lambda}, \hbar, \wp_{Ca}, \tau_{xCa}$ Electrical numerical parameters and constants

$\overset{0}{\breve{V}}, \breve{V}$ Initial and current intrauterine volumes

$k_{(\pm)i}$ Rate constants of chemical reactions

A, B, C, D   Matrices of rate coefficients

$\mathbf{X}(X_i)^T, \mathbf{C_0}(C_i)^T$ Vectors of reacting substrates

Abbreviations

AC	Adenylyl cyclase
ACh	Acetylcholine
AD	Adrenaline
ATP	Adenosine-5′-triphosphate
BK_{Ca}	Large conductance Ca^{2+} activated K^+ channel
cAMP	Cyclic adenosine monophosphate
COX-1/2	Cyclooxygenase 1 and 2
CP	Cytoskeletal proteins
DAG	Diacylglycerol
DNA	Deoxyribonucleic acid
ECM	Extracellular matrix
$EP_{1,2,3A,D}$	Prostaglandin $E_{1,2,3A,D}$ receptors
FAP	Focal adhesion proteins
hCG	Human chorionic gonadotrophin
IL	Interleukin
IP_3	Inositol-1,4,5-triphosphate
K_{ATP}	ATP-sensitive K^+ channel
K_V	Voltage-gated K^+ channel
MAPK	Mitogen-activated protein kinase
MLCK	Myosin light chain kinase
*m*PR, *n*PR	Membrane and nuclear progesterone receptors
mRNA	Messenger ribonucleic acid
MRI	Magnetic resonance imaging
OT	Oxytocin
P2X	Purine type 2X receptors
$PGF_{2\alpha}$	Prostaglandin $F_{2\alpha}$
PKA	Protein kinase A

PKC	Protein kinase C
PLC	Phospholipase C
PR	Progesterone
SK_{Ca}	Small conductance Ca^{2+} activated K^+ channel
SR	Sarcoplasmic reticulum

Chapter 1
Biological Preliminaries

Facts are the air of science. Without them a man of science can never rise.

I. Pavlov

1.1 The Uterus

The nonpregnant human uterus is a hollow, thick-walled, organ situated deeply in the pelvic cavity. It measures on average 7.5 cm in length, 5 cm in breadth, at its upper part, and nearly 2.5–4 cm in thickness (Fig. 1.1). Anatomically the organ is divided into: (1) the fundus, (2) the body, (3) the uterotubal angles, and (4) the cervix. A region between the body and the cervix is call the isthmus. The cervix of the uterus is connected to the vagina. It is conical or cylindrical in shape, with the truncated apex.

The uterus is supported by eight ligaments: anterior, posterior, and dual lateral, uterosacral, and round ligaments. The anterior and posterior ligaments consist of the vesicouterine and the rectovaginal folds of the peritoneum. They contain a considerable amount of fibrous tissue and nonstriped muscular fibers. At one end, they are attached to the sacrum and constitute the uterosacral ligaments. The two broad ligaments pass from the sides of the uterus to the lateral walls of the pelvis. The round ligaments are two flattened bands situated between the layers of the broad ligament in front of and below the uterine tubes. They consist principally of muscular tissue, prolonged from the uterus, and of some fibrous tissue. The ligaments contain blood and lymph vessels, and nerves. Additionally, there are fibrous tissue bands on either side of the cervix uteri – the ligamentum transversalis coli. They are attached to the side of the cervix uteri, to the lateral walls of the pelvis and to a part of the vagina.

The arterial supply to the uterus is mainly from the hypogastric artery. The arteries are remarkable for their tortuous course and frequent anastomoses within the wall of the organ. The venous return corresponds to the course of the arteries

R.N. Miftahof and H.G. Nam, *Biomechanics of the Gravid Human Uterus*,
DOI 10.1007/978-3-642-21473-8_1, © Springer-Verlag Berlin Heidelberg 2011

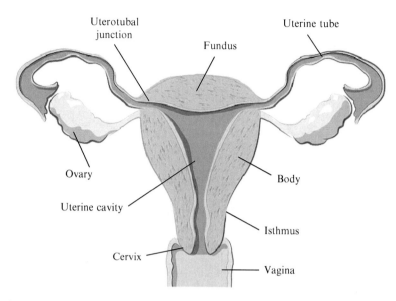

Fig. 1.1 Anatomy of the nonpregnant human uterus

and drains in the uterine plexi. In the pregnant uterus the arteries carry the blood to, and the veins convey it away from the intervillous space of the placenta.

The nerves are derived from the hypogastric and ovarian plexi, and from the third and fourth sacral nerves. It suffices to note that there is afferent and no efferent innervation to the uterus.

The actual position of the body of the uterus in the adult is liable to considerable variation. It depends chiefly on the condition of the nearby organs – the bladder and rectum. When the bladder is empty, the uterus is bent on itself at the junction of the body and cervix and lies upon the bladder. As the bladder fills, it becomes more erect, until with a fully distended bladder, the fundus of the uterus may face the sacrum. Movements of the cervix though are very restricted.

The cavity of the nonpregnant womb resembles a mere flattened slit. It is triangular in shape, with the base being formed by the internal surface of the fundus and the apex – by the internal orifice of the cervix. The uterine tubes connected on either side of the fundus allow the ova to enter the uterine cavity. If an ovum is fertilized, it imbeds itself in the uterine wall and is normally retained there until prenatal development is completed.

1.2 Functional Unit

The human uterus is a multi-component system with optimal spatiotemporal arrangements among morphological elements. It is composed of three histologically distinct layers (from inside out): the endometrium, the myometrium, and the perimetrium. The endometrium of the womb is lined with a surface epithelium and has three types of cells: secretory, ciliated, and basal. The main function of the endometrium is to provide implantation of a fertilized ovum and to support the growth of a fetus. The uttermost perimetrium is composed of a thin layer of connective tissue – collagen and elastin fibers.

The most prominent layer – the myometrium – is divided into three poorly delineated self-embedded layers (strata): the strata supravasculare, vasculare, and subvasculare. The morphostructural functional unit of the muscle tissue of the uterus is the uterine smooth muscle cell – myocyte. It has a characteristic spindle-like shape which is determined by the cytoplasmic cytoskeleton (Yu and Bernal 1998; Small and Gimona 1998). The latter is formed by intracellular thin actin, ~6 nm, intermediate, ~10 nm, and microtubular filaments, ~20–25 nm, in diameter. Arranged into a lattice, with multiple insertions of cytoplasmic dense bodies, and attached to the cell membrane at the sites of plasmalemmal dense plaques they guarantee the integrity, strength, and high degree of deformability of the myometrium.

Plasmalemmal dense plaques contain focal adhesion complexes which comprised multifunctional proteins, i.e., integrins, syndecans, paxillin, vincullin, tallin, and a family of kinases, i.e., focal adhesion, extracellular signaling, and c-Src kinases (Gerthoffer and Gunst 2001; MacIntyre et al. 2008; Li et al. 2003). They make direct structural and functional contacts between the intracellular cytoskeleton and the extracellular matrix (ECM), and act as mechanosensors in tyrosine kinase signaling pathways (Gabella 1984; Burridge and Chrzanowska-Wodnicka 1996; MacPhee and Lye 2000; Williams et al. 2005).

Individual myocytes are integrated into bundles, $\simeq 300 \pm 100$ μm, surrounded by fine fibrillar matrix interspersed with microvasculature. Multiple pore structures between myometrial cells allow cell-to-cell communication via diffusible intracellular components – gap junctions. Immunofluorescent studies showed that they are formed predominantly by the connexin-43 protein with varying degrees of connexins 40 and 45 (Sakai et al. 1992; Tabb et al. 1992; Kilarski et al. 2001). Gap junctions provide the structural basis for electrical and metabolic communications, support synchronization and long range integration in myocytes. These ensure the property of myogenic syncytia necessary for the coordinated, phasic contractions of labor. Immunohistochemical labeling studies has demonstrated that the expression of connexin-43 is not even in the pregnant uterus. There is a significant increase in the protein content in the fundus compared to the lower segment of the organ. This fact has been suggested to be pivotal in electrogenic coupling asymmetry and conductance anisotropy (Blanks et al. 2007). Bundles of myocytes are further organized into fasciculi, $\simeq 1$–2 mm in diameter, which are interconnected into a three-dimensional network. The network is submerged into ECM that provides a supporting stroma to myocytes (Young and Hession 1999; Blanks et al. 2007).

Myometrial cells acutely adjust their structure and function by reorganizing the cytoskeleton and altering signaling pathways (Salomonis et al. 2005). Thus, hypertrophic changes in pregnant uterus are associated with remodeling of ECM from

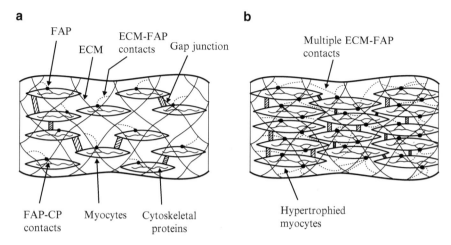

a

FAP
ECM
ECM-FAP contacts
Gap junction

FAP-CP contacts
Myocytes
Cytoskeletal proteins

b

Multiple ECM-FAP contacts

Hypertrophied myocytes

Fig. 1.2 Structure of myometrium in a nonpregnant (**a**) and pregnant (**b**) uteri

a fibrillar to sponge-like architecture (Gunja-Smith and Woessner 1985; Leppert and Yu 1991; Metaxa-Mariatou et al. 2002). Activation of integrin and fibrionectin proteins promotes the development of new cell-matrix contacts and formation of the mesh-like fine fibrillar matrix on the cell surface and their cytoplasmic domains attach to cytoskeletal proteins (CP) (Williams et al. 2005; Robinson et al. 2004; Woodward et al. 2001; Shynlova et al. 2007) (Fig. 1.2). Such changes ensure myometrial homogeneity and affirm even stress–strain distribution in the tissue.

The contractile apparatus of myocytes consists of thin-actin and thick-myosin filaments, a family of special proteins and kinases, e.g., light chain myosin, tropomyosin, calmodulin, h-caldesmon, calponin, myosin light chain kinase, and myosin phosphatase. Actin filaments are single helical coils of actin associated with tropomyosin and caldesmon. Myosin filaments are made out of two coil rod-like structure heavy chains with a globular head domain. A principal determinant of the dynamics of contractions is free cytosolic calcium (Ca_i^{2+}) that triggers the cyclic actin–myosin complex formation. Two types of contractions – tonic and phasic – are produced by the myometrium. Thus, in the late phase of labor and immediately after delivery the myometrium produces tonic contractions. During delivery, though, it undergoes intensive phasic contractions.

1.3 Electrophysiological Properties

Myometrial contractions are driven by waves of electrical activity. The electrical repertoire of myocytes depends on balanced function of plasma-lemmal ion channels. Composed of discrete transmembrane proteins, they allow for the selective transfer of ions across the cell membrane. L- and T-type Ca^{2+},Ca^{2+}-activated

K^+, voltage-dependent K^+ and Na^+, and Cl^- channels reflect the multiplicity and complexity of mechanisms involved in the regulation of uterine excitability and contractility. The presence of L-type Ca^{2+} channels in the human uterus has been confirmed by electrophysiological, pharmacological, and molecular studies (Young et al. 1993; Parkington and Coleman 2001; Chien et al. 1996; Collins et al. 1996; Mershon et al. 1994). They are formed of five distinct subunits: α_1, α_2, β, δ, and γ. The α_1-subunit contains the channel pore, voltage sensor and drug binding sites, while α_2, β, δ and γ-subunits modulate the channel's activity. L-type channels possess characteristics of long-lasting, high-voltage activated channels. They ensure the main influx of extracellular calcium ions, Ca_0^{2+}, during depolarization.

Three subfamilies of T-type Ca^{2+} channels, which differ in α-subunits, have been identified by cloning technique in human myocytes. They have distinct activation/deactivation kinetics: they are activated at low voltage and remain open for a short period of time (Young and Zhang 2005; Monteil et al. 2000; Lee et al. 1999). Experimental data suggest that T-type channels are responsible for the generation of spikes and pacemaker activity, and play a key role in regulation of the frequency of phasic contractions (Lee et al. 2009a, b).

Potassium channels constitute a superfamily of five channels: the large Ca^{2+}-activated K^+ (BK_{Ca}), intermediate (IK_{Ca}), small conductance (SK_{Ca}), voltage-gated (K_v), and ATP-sensitive (K_{ATP}) potassium channels. The channels are associated with caveolin-1, lipid rafts, and scaffolding proteins. They are located in invaginations of the cell membrane – caveolae (Brainard et al. 2005). The BK_{Ca} channel is made of six transmembrane proteins. The two distinct α- and β-subunits regulate channel's sensitivity to Ca^{2+}. They remain uncoupled at low $[Ca_i^{2+}]$ and switch to a calcium sensor mode with the rise in intracellular calcium. It has been proposed that the up-regulation of BK_{Ca} channels during pregnancy is the major factor that sustains quiescence and relaxation of the uterus (Korovkina et al. 2006).

Two types of K_v channels are identified in human myocytes: delayed rectifying and rapidly inactivating. They are formed by a single unit of six transmembrane proteins and the pore-hairpin loop (Jan and Jan 1994). Electrophysiological studies confirmed that channels are responsible for repolarization and slow wave generation in myometrium.

Data on kinetics and dynamics of K_{ATP} and SK_{Ca} channels are limited. There is no compelling experimental evidence to support their existence in the human myometrium. Even if they are present, their density is expected to be relatively low. Their overall effect is the maintenance of the resting membrane potential at a constant level and inhibition of phasic uterine contractions. Modulation and function of these channels during pregnancy has not been fully studied.

The expression of the two voltage-gated Na^+ channels has been reported in the human uterus (Young and Henrdon-Smith 1991). Their role in gestation and labor is still a matter of debate. It has been proposed that the current facilitates and augments the propagation of the wave of excitation and synchronizes contractions.

The distinct role of Cl^- channels is unclear due to the uncertainty of their molecular identity (Adaikan and Adebiyi 2005; Hartzell et al. 2005). Calcium-activated chloride currents have been recorded on the isolated uterine smooth muscle cells. They are evoked by elevation of Ca_i^{2+} and have distinctive biophysical properties (Ledoux et al. 2005). However, until the molecular structure of these channels is resolved, we can only speculate on their possible functional role.

The resting membrane potential of freshly isolated human uterine smooth muscle cells varies between $V^r \simeq -70 \div -45\,mV$. The level depends on their gestational status, i.e., cells become more depolarized toward the full term and delivery. Direct measurements revealed that the specific membrane resistance of myocytes is $R_m \simeq 6\,k\Omega\,cm^2$ and the capacitance is $C_m \simeq 1.6\,\mu F/cm^2$ (Pressman et al. 1988; Inoue et al. 1990; Parkington et al. 1999).

The myometrium is spontaneously excitable and produces low amplitude, ~10–20 mV, and frequency, ~0.008–0.1 Hz, slow waves and high frequency, ~17–25 Hz, and amplitude, ~40–50 mV, action potentials (spikes). Although the existence of slow waves in the myometrium is being debated, in vivo and in vitro recordings from pregnant human uteri clearly demonstrate long-period, ~1.5–2 min, oscillations of membrane potentials (Garfiled and Maner 2007; Kuriyama et al. 1998; Blanks et al. 2007; Young 2007; Tribe 2001). The dynamics of action potentials is greatly influenced by neurohormonal factors. Spikes are almost absent during the entire pregnancy and are generated only at late-term, labor, and delivery. Their duration varies between 100 and 250 ms. The velocity of the propagation of excitation in the myometrium is 1.0–9.0 cm/s and depends on the anatomical site and physiological status of the uterus (Kryzhanovskaya-Kaplun and Martynshin 1974; Kao and McCullough 1975). The excitatory waves synchronized myoelectrical activity in the organ and organize contractions. Noninvasive myographic and electrohysterographic studies revealed the preferred propagation of excitation from the fundus toward the cervix in pregnant human uteri (Planes et al. 1984; Horoba et al. 2001; Garfield et al. 2005; Ramon et al. 2005a, b; Marque et al. 2007).

Pacemaker cells have not been found in the uterus yet. Ultrastructural and immunohistochemical studies have helped to identify interstitial Cajal-like cells in the myometrium and fallopian tubes (Popescu et al. 2007). This is based mainly on morphological similarities with cells of Cajal found in the gastrointestinal tract, where their role as pacemakers has been "firmly" established. Extrapolations have been made to associate them with uterine pacemaker cells. In contrast, electrophysiological recordings of spontaneous spiking activity from impaled myocytes suggested the concept of variable pacemakers, in other words any myometrial muscle cell or a group of cells is capable of acting as either a pacemaker or pacefollower (Parkington and Coleman 2001).

1.4 Neuroendocrine Modulators

Although electrical waves of depolarization induce spontaneous contractions of the myometrium, fine-tuning of its mechanical reactivity during gestation and parturition is provided by the fetal–maternal hypothalamic–pituitary–adrenal–placental axis. This involves a number of multiple neuropeptides and hormones. Their effects are spatiotemporally integrated at the synaptic level which quantifies a required amount of chemically coded information that is to be passed into the cell. The concept of neuroendocrine modulation of uterine activity includes intricate signal transduction mechanisms of synthesis, storage, release and utilization of transmitters, their binding to and activation of receptors and enzymatic degradation with subsequent initiation of a variety of cellular events.

Neuroendocrine modulators are broadly divided into contractors (utero-tonics), among which the most powerful are oxytocin (OT), prostaglandins $F_{2\alpha}$ and E_2 ($PGF_{2\alpha}$, PGE_2), and acetylcholine (ACh), and relaxants (tocolytics) – progesterone (PR) and adrenaline (AD). They exert their effects through a number of specific and G-protein-coupled receptors (Sanborn et al. 1998a, b). The expression and sensitivity of receptors depend on the gestational status. Moreover, the receptor co-localization allows the myocyte to respond to more than one neuroendocrine modulator at a time and thus, to assure its highly adaptive properties.

Acetylcholine is the ubiquitous neurotransmitter in the central and autonomic nervous systems. It is synthesized in neurons by the enzyme choline acetyltransferase from choline and acetyl-CoA. ACh molecules are stored in vesicular form in the presynaptic nerve terminals. The main trigger for its release is cytosolic calcium, Ca_i^{2+}, which is increased during depolarization of the nerve terminal. Free Ca_i^{2+} binds to reactive Ca^{2+} centers on the vesicles, causes their fusion with the presynaptic cell membrane, and vesicular ACh exocytosis into the synaptic cleft. The process has a quantum character. The excess free fraction of ACh is converted by acetylcholinesterase enzyme into the inactive metabolites, choline and acetate. The enzyme is abundant in the synaptic cleft, and rapid clearing of the excess of ACh is essential for proper muscle function. Another part diffuses to the postsynaptic membrane on the myocyte and binds to muscarinic type 2 (μ_2) $G\alpha_{i(1-3)}$ coupled receptors (Nakanishi and Wood 1971; Caulfield and Birdsall 1998). The ACh-receptor complex generates the postsynaptic response, i.e., the inhibition of the intracellular cAMP pathway. It involves the inactivation of adenylyl cyclase (AC) enzyme, monophosphate (cAMP) production from adenosine $5'$-triphosphate (ATP), and the protein kinase A (PKA) pathway (Moss and Getton 2001). As a result there is a suppression of phosphorylation of myosin light chain kinase (MLKC) and potentiation of myometrial smooth muscle contractility. cAMP decomposition into adenosine monophosphate (AMP) is catalyzed by the enzyme phosphodiesterase 4-2B (PDE4-2B).

The neurohypophysial hormone oxytocin is a potent and specific stimulant of uterine contractions. The peptide is synthesized as an inactive precursor protein from the OT gene, which is progressively hydrolyzed by a series of enzymes to form the active oxytocin nonapeptide. OT is packaged in large, dense-core vesicles of the corpus luteum and placental cells and is secreted upon stimulation by exocytosis. Myometrial OT receptors are functionally coupled to $G\alpha_{q/11}$ proteins. Their expression and sensitivity to OT increases 12-fold toward late pregnancy and during labor (Rezapour et al. 1996). Activation of the receptor together with $G\beta\gamma$ stimulates the phospholipase C (PLC) pathway. Additionally, OT activates the c-Src and focal adhesion kinases which adds to the production of inositol-1,4,5-triphosphate (Ilic et al. 1997; Yamamoto and Miyamoto 1995). It has been demonstrated experimentally that OT directly increases the calcium influx through voltage-gated or ligand-operated channels. The effect is to be nifedipine-insensitive (Sanborn et al. 1998a, b). Oxytocin is metabolized predominantly in placenta and to a lesser extent in myometrium and deciduas by two major enzymes: aminopeptidase and postproline endopeptidase.

Prostaglandins $F_{2\alpha}$ and E_2 are prostanoids and belong to the eicosanoid family of biologically active lipids. They are synthesized in the decidua, the chorion leave and the amnion from arachodinic acid by a combination of cyclooxygenases (COX-1 and 2) and specific synthase enzymes (Coleman et al. 1994). PG levels depend on the presence and activity of the principle metabolizing enzyme, 15-hydroxy-prostaglandin dehydrogenase. Released in the bloodstream, they exert their paracrine effect on the myometrium. It is accepted that $PGF_{2\alpha}$ mediates its actions via FP-$G\alpha_{q/11}$-protein-coupled receptors and activation of the PLC pathway (Olson et al. 2003; Hertelendy and Zakar 2004). Also experiments on isolated myometrial cells have led to the notion that its primary effect is to promote Ca^{2+} entry through L-type Ca^{2+} channels (Molnar and Hertelendy 1990; Anwer and Sanborn 1989). PGE_2 transmits signals in human myometrium by binding to the four distinct $G\alpha_i$-coupled receptors ($EP_{1-3A,D}$). EP_1- and EP_{3D}-type receptors cause an elevation in intracellular calcium and increase contractility via activating L-type Ca^{2+} channels, whereas the stimulation of EP_2 receptor increases cAMP production. Activation of EP_{3A} augments mitogen-activated protein kinase (MAPK) – serine/threonine-specific protein kinases – activity and impedes the cAMP pathway. Additionally, the stimulation of EP_{3D} receptor enhances the phosphatidyl-inositol turnover via the PLC pathway and, thus adds to the rise in cellular calcium (Asboth et al. 1997; Kotani et al. 2000). Prostaglandins are deactivated in placenta by 15-hydroxyprostaglandin dehydrogenase and 13,14-prostaglandin reductase.

Adrenaline (AD), also known as epinephrine, is a hormone and a neurotransmitter. It is synthesized in the cytosol of adrenergic neurons and adrenal gland medullary cells. The main precursor – the amino acid tyrosine – is converted in an enzymatic pathway into a series of intermediates namely L-dopa, dopamine, noradrenaline, and ultimately adrenaline. Newly synthesized molecules are stored in chromaffin granules. Intracellular rise of Ca^{2+} triggers its release by exocytosis. AD acts via two main groups of G-protein-coupled adrenoceptors, α and β. Subtypes α_1, α_2, β_1, β_2, and β_3 adrenoceptors ($R_{\alpha i,\beta i}$) are present in the human uterus. The excitatory α_1-AR is coupled to $G\alpha_{q/11}$ protein and upon activation stimulates the PLC signaling pathway. The α_2-$R_{\alpha 2}$ is linked to $G\alpha_i$ protein and decreases cAMP production. β_1 and β_3 $R_{\beta i}$ are coupled to $G\alpha_s$, and β_2 type to $G\alpha_s$ and $G\alpha_{i(1-3)}$ proteins. Their function is associated with activation of the cAMP-dependent pathway. Adrenaline exerts negative feedback to down-regulate its own synthesis at the presynaptic α_2 adrenoceptors. Excess adrenalin is removed by two mechanisms: uptake-1 and uptake-2. The uptake-1 mechanism involves deamination of AD by monoamine oxidase, and the uptake-2 mechanism its degradation by catechol-O-methyltransferase enzyme to metabolic products.

Progesterone belongs to a class of hormones called progestogens, and is the major naturally occurring human progestogen. It is synthesized from cholesterol to form pregnenolone which is further converted to progesterone in the presence of 3β-hydroxysteroid dehydrogenase/$\Delta(5)$–$\Delta(4)$ isomerase. PR exerts its action through the G-protein-bound membrane (mPR), and nuclear (nPR) ligand-activated transcription factor receptors (Roh et al. 1999; Karteris et al. 2006; Chapman et al. 2006). Three isoforms, PR-A, PR-B and PR-C, that differ in their molecular weight, affinity and functionality have been identified in the human uterus. The mPRs are coupled to $G\alpha_i$ proteins and their activation results in a decline in cAMP levels. Activation of mPR receptors leads to transactivation of nPR-B. The PR–nPR-B receptor complex undergoes dimerization and enters the nucleus where it binds to DNA. The following transcription leads to formation of messenger ribonucleic acid (mRNA) and the production of specific proteins. 17α- and 21α hydroxylases are responsible for progesterones enzymatic conversion to the mineral corticoid, cortisol, and androstenedione.

There is accumulating evidence that inflammatory cytokines, IL-1β, IL-6 and IL-8 and tumor necrosis factor are involved in normal term labor. For example, IL-1β is produced by macrophages, monocytes, fibroblast, and dendritic cells as a proprotein. It is proteolytically converted to its active form by caspase 1. This cytokine is a powerful pro-labor mediator through the stimulation of the PLC pathway, the induction of Ca^{2+} ion release from the sarcoplasmic reticulum, the activation of p38MAPK, the enhanced production of COX-2 enzyme and prostaglandins. IL-8 expression in the myometrium is maximal at term (Elliott et al. 2000). The chemokine contributes to cervical maturation by stimulating extravasation of neutrophils. The latter release matrix metalloproteinases that denature the collagen within ECM increases cervical compliance and uterine contractility (Osmers et al. 1995a, b).

The list of potential uretotonics can be extended to include estrogen, tachykinins, adenosine 5′-triphosphate, endothelin-1, platelet-activating factor A$_2$, thrombin, as well as tocolytics – human chorionic gonagotrophin (hCG), relaxin, calcitonin-gene-related peptide. It is apparent though that electrical, neuroendocrine, and mechanical stimuli can initiate and regulate mechanical activity in pregnant myometrium at term. However, operating alone or conjointly they cannot sustain the required strength, frequency and duration of contractions which are needed to expel the products of conception. To be robust they must be organized and integrated in time and space. Failure to achieve dynamic coordination results in pathological conditions such as dystocia and impediment of labor.

1.5 Coupling Phenomenon

Individual myocytes require the presence of flexible dynamic links among intrinsic electrical, chemical (neurohormonal), and mechanical processes to perform as a physiological entity. These complex forward-feedback interactions modulate the function of extra- and intracellular protein pathways, expression and distribution of gap junctions, ion channels, surface membrane and nuclear receptors in the organ.

Electro-chemo-mechanical coupling is a sequence of events heralded by myometrial contractions that are preceded by the wave of depolarization of the cell membrane and/or ligand–receptor complex formation. The main functional link in the cascade of processes is intracellular free calcium. Two major sources for it are: (1) the flux of extracellular calcium inside the cell, and (2) its release from the internal store – sarcoplasmic reticulum (SR). The influx is ensued by voltage-gated L- and T-type, and ligand-operated L-type Ca^{2+} channels. Depolarization as well as binding of OT, $PGF_{2\alpha}$ and PGE_2 to the membrane receptors increase channels' open state probability, with resulting rise in $[Ca_i^{2+}]$. Recent studies have highlighted a possible direct activation of L-type Ca^{2+} channels by DAG; however, the exact mechanism is unclear. The blockade of L-type Ca^{2+} channels with selective antagonists, nifedipine and verapamil, abolishes cytosolic calcium transient and reduces both spontaneous and induced mechanical activity (Lee et al. 2009a, b). The release of Ca^{2+} from the SR is accomplished by stimulation of the ryanodine and IP_3 receptors on its membrane. However, the amount of Ca_i^{2+} released is small compared to that entering the cell through Ca^{2+} channels (Taggart and Wray 1998; Kiputtayanant et al. 2002). Free cytosolic calcium binds to calmodulin protein to form the Ca^{2+}–calmodulin complex. It further induces the cascade of downstream reactions with subsequent inhibition in phosphorylation of MLCK, and activation of h-caldesmon, calponin, and the light chain myosin. Active light myosin interacts with actin to form actin–myosin crossbridges and, hence, contraction.

Experiments with wortmannin, a myosin light chain kinase inhibitor, demonstrated that myocytes continued to produce the active force in the presence of the drug although of lower strength and intensity. It has been suggested that other intracellular pathways dependent on the Ca_0^{2+} influx, are involved in myometrial contractility (Longbottom et al. 2000). Thus, membrane dense plaque proteins serve as molecular signaling platforms to mechano-chemical coupling. An externally applied stretch induces phosphorylation of tyrosine-dependent focal adhesion proteins, c-Src and intracellular kinases and promotes interaction of structural proteins, mainly paxillin, vincullin and talin. The formation of a focal adhesion proteins (FAP) complex and its association with phosphorylated h-caldesmon leads to contraction (Wu et al. 2008; Li et al. 2007, 2009a, b, c; Gerthoffer and Gunst 2001). It is noteworthy that tyrosine kinase directly modulates the L-type Ca^{2+} channel (Shlykov and Sanborn 2004).

The actin–myosin complex formation is universally accepted as the only mechanism responsible for force development by smooth muscle. However, in vitro studies have recorded active contractions of the myometrial extracellular matrix (Fitzgibbon et al. 2009). Although the exact mechanism remains unclear, actin polymerization and its remodeling have been offered as an explanation of the contractility (Smith et al. 1997; Wu et al. 2008).

Relaxation of the myometrium is brought about by the processes of hyperpolarization, reduction in $[Ca_i^{2+}]$, and activation of the cAMP-dependent inhibitory pathway. Activation of BK_{Ca}, K_v, and Cl^- channels is the major regulatory feedback element that causes hyperpolarization of the cell membrane. Two mechanisms, extrusion by the Na^+/Ca^{2+} exchanger and active re-uptake of Ca_i^{2+} into the SR, control cytosolic calcium at required levels (Adaikan and Adebiyi 2005; Khan et al. 2001; Kiputtayanant et al. 2002; Price and Bernal 2001; Yuan and Bernal 2007). Progesterone and adrenaline stimulate PKA via the cAMP-dependent pathway. PKA further phosphorylates MLCK and hence, attenuates its affinity for the Ca^{2+}–calmodulin complex resulting, finally, in dissociation of the actin–myosin complexes. The protein also promotes desensitization of $G\alpha^q$ receptors, decreases gap junction permeability, and inhibits the phospholipase C pathway with a cumulative effect of myometrial relaxation (Price and Bernal 2001; Yuan and Bernal 2007).

1.6 Crosstalk Phenomena

Over the years, the physiological function of different modulators have been studied and analysed independently. The important factor that has been overlooked though is the cotransmission by multiple signaling molecules and their integrated effect on myometrial contractility. The positive crosstalk between caveolae co-localized β_2-ARs, BK_{Ca}, and K_v channels facilitates hyperpolarization of the membrane and uterine relaxation (Hamada et al. 1994; Chanrachakul et al. 2004; Sanborn et al. 1998a, b, 2005). Activation of β adrenoceptors was also shown to exert a negative regulation on the PLC pathway by reducing IP_3 production through the cAMP-dependent mechanism (Khac et al. 1996). Co-localization of α_1-adrenoceptors with the OT receptor permits the adrenaline modulation of myometrial activity (Liu et al. 1998; Gimpl and Fahrenholz 2001; Hamada et al. 1994). However, the number of α_1- and β_1-ARs in human myometrium is relatively small and thus, their role in relaxation is insignificant.

Immunoprecipitation and matrix-assisted laser desorption "time-of-flight" mass spectrometry studies revealed strong association of BK_{Ca} channels with α- and γ-actin cytoskeletal filaments. It was postulated that mechanical stretch can modulate channel activity and vice versa. Stretch of uterine myocytes has also been shown to increase the expression of contraction-associated proteins, the up-regulation of COX-2 mRNA, p38MAPK, and the chemokine IL-8 (Sooranna et al. 2004; Oldenhof et al. 2002; Loudon et al. 2004). The induction of labor by prostaglandins is associated with greater expression of IL-8, suggesting an interaction between PRs and the chemokine (Chan et al. 2002).

Although OT and PGE_2 stimulate IP_3 accumulation and Ca^{2+} release from the SR via the IP_3-sensitive mechanism, only OT-induced high-frequency Ca^{2+} oscillations are ryanodine-sensitive (Asboth et al. 1996; Burghardt et al. 1999). Inhibition of the $G\beta\gamma$ subunit stimulates $PLC\beta2$ and $PLC\beta3$ through phosphorylation of the cAMP-dependent PKA indicating possible combined regulatory effects of acetylcholine, adrenaline, and progesterone (Yue et al. 1998; Engstrøm et al. 1999; Mhaouty-Kodja et al. 2004). The secretion of PGE_2 is controlled by human chorionic gonadotrophin through expression of COX-2 messenger ribonucleic acid (Zhou et al. 1999). The stimulant effects of $PGF_{2\alpha}$ on uterine contractility are also augmented by ATP via activation of $P2X_{1,3}$ receptors (Ziganshin et al. 2005, 2006). Chemokines have shown to increase (IL-1β) or decrease (IL-6) $PGF_{2\alpha}$ levels in cultures of human uterine myocytes (Liang et al. 2008). The effects are mediated in part by protein kinase C, but are independent of MAPK, PLC and IP_3 kinases. Interestingly, mechanical stretch has no effect on $PGF_{2\alpha}$ mRNA expression (Sooranna et al. 2005).

Extensive crosstalk exists among progesterone receptors. PR-B receptors are antagonized by PR-A receptors which are dominant repressors of transcription. Therefore, it has been suggested that an increase in their expression during labor causes "functional" progesterone withdrawal and induction of delivery (Pieber et al. 2001). nPR-C receptors reside mainly in the cytosol of myocytes and have high affinity for progesterone. Their activation diminishes the concentration of PR–nPR-B complexes and, thus promotes contractions. Additionally, progesterone represses connexin-43 levels and gap junction density in myocytes at term (Ambrus and Rao 1994). Hence, PR can attenuate electrical coupling and disrupts the process of propagation of the wave of excitation in myometrium.

Human chorionic gonadotrophin has an essential role in the establishment of pregnancy and its continuation. There is convincing experimental evidence that it promotes uterine quiescence by binding positively to hCG-$G\alpha_s$ protein-coupled receptors and stimulating the cAMP/PKA pathway. Currently, it is not clear what specific targets are phosphorylated downstream by PKA. The available data suggest though that the hormone interferes with OT signaling (Eta et al. 1994). Moreover, hCG down-regulates gap junction formation, directly activates BK_{Ca} channels, impedes Ca_i^{2+} availability by inactivation of voltage-gated Ca_i^{2+} channels, and inhibits PDE5 enzyme (Doheny et al. 2003; Lin et al. 2003; Ambrus and Rao 1994; Belmonte et al. 2005).

Net myometrial contractile activity is determined by the balance of receptors present. To date, no detailed quantitative studies have been conducted to analyze the pattern of the receptor distribution and receptor types in human uteri. It is of utmost importance though to establish receptor topography and their expression throughout the fundus to cervix, during gestation and labor. Ligand binding studies revealed heterogeneous ratios of EP_1, EP_3, and $PGF_{2\alpha}$ receptors in the organ with the highest concentration in the fundus (Giannopoulos et al. 1985). The lower segment of the uterus at term has been demonstrated to respond more to relaxatory (EP_2) rather than to excitatory (EP_3) receptor activation (Brodt-Eppley and Myatt 1999). Myometrial oxytocin receptor levels have been found in greater abundance in the fundus and the corpus than in the lower uterine segment where their concentration was the lowest (Fuchs et al. 1984).

Functional binding, Western blot, and polymerase chain reaction experiments showed the predominance of β_3-adrenoceptors compared with the β_2-AR type in myometrium (Rouget et al. 2005; Bardou et al. 2007). There is also a gradual increase in the total number of α_1-ARs during pregnancy (Dahle et al. 1993). Little is known though about their hormonal modulation. Elevation of 17β-estradiol has been demonstrated to increase the myometrial response to a selective β_3-AR agonist (Ferre et al. 1984). Tachykinin NK_2, purinergic $P2X_1$, serotonergic $5HT_{2A}$, and histamine H_2 receptors have been identified recently in human pregnant myometrium (Patak et al. 2000, 2003; Ziganshin et al. 2006; Willets et al. 2008; Cordeaux et al. 2009). However, data about their distribution, co-localization, and co-transmission are currently lacking.

Exercises

1. The concept of a functional unit has proven to be successful and robust in studying various organs and systems in the body. What are the general principles for the construction of a functional unit?
2. What experimental evidence supports the approximation of a morphologically heterogeneous myometrium as a mechanically homogeneous syncytium?
3. Slow wave electrical activity is an intrinsic phenomenon and has been observed in smooth muscle cells of the small intestine, colon, trachea, etc. It provides stability for the propagation of excitation and the coordination of motility. What are physiological bases for oscillatory activity in the human myometrium?.
4. The question of the "pacemaker", its morphological identity, electrophysiological properties and function has always captivated biologists. Pacemaker cells have not yet been found in the human uterus, although experiments and clinical data demonstrate that the excitation normally originates at the uterotubal junctions. What are the possible underlying mechanisms of pacemaker activity?
5. Little is known about the distribution of receptors and receptor types within the uterus. Recent experimental data suggest pregnancy-dependent expression and sensitivity of receptors in different anatomical sites of the organ. Construct the map for oxytocin (OT) and prostaglandin $F_{2\alpha}$ and E_i ($PGF_{2\alpha}$ and PGE_i) receptor distribution in the pregnant human uterus at term. Discuss the possible role of OT and PGs in triggering pacemaker activity.
6. It has been emphasized that drug selectivity and specificity are the key in the treatment of a disease. In contrast with the theory, "dirty" drugs with low receptor selectivity and specificity are more clinically beneficial. Is specificity important in the control of labor? (*Hint*: analyze the role of different anatomical sites of the uterus in the biomechanics of labor).
7. Application of the double balloon device in combination with intravaginal placement of PGE_2 tablets is the procedure used to induce labor. Provide biomechanical explanation for the rational of the procedure.
8. Wortmannin, a myosin light chain kinase inhibitor, does not abolish contractility of myocytes. However, they continue to generate active forces of low strength. Discuss the mechanisms (pathways) involved in mechanical activity of myometrium.

9. Myometrium, an electrogenic syncytium, possesses cable and oscillatory characteristics. The former sustains the spread of the wave of depolarization in the organ, the latter its contractility. Identify morphostructural elements responsible for these properties.

10. Results of a reductionist approach in search for the channel of myometrial quiescence suggest BK_{Ca} channels as the key factor. Are the BK_{Ca} channels indeed the channels of myometrial "quiescence"?

Chapter 2
Models of the Gravid Uterus

Mathematics as an expression of the human mind reflects the active will, the contemplative reason, and the desire for aesthetic perfection.

R. Courant

2.1 Biological Changes in Pregnant Uterus

During pregnancy the uterus evolves considerably with dynamic changes related to both special and temporal processes – a process that is closely controlled by intrinsic and extrinsic regulatory mechanisms. The gravid organ extends from the pelvis and occupies the lower and middle abdomen. It undergoes changes in size and structure to accommodate itself to the needs of the growing embryo – "uterine conversion" (Reynolds 1949). The pregnant human uterus exhibits marked differences in morphology and protein expression patterns compared to the non-pregnant organ collectively referred to as "phenotypic modulation". The process includes four distinct phenotypes (phases): (1) early proliferative, (2) intermediate synthetic, (3) contractile phase, and (4) highly active labor.

During the first stage, uterine myocytes proliferate rapidly, predominantly in the longitudinal layer. The following synthetic phase is remarkable for myometrial cell hypertrophy associated with increased synthesis and deposition of interstitial matrix. With progression of pregnancy, the uterus advances into a contractile phenotype. At this stage, the rate of cellular hypertrophy remains constant. There is a continuous increase in the γ-smooth muscle actin, expression of L-, T-type Ca^{2+}, and Ca^{2+}-activated K^+ ion channels, receptors for oxytocin and prostaglandins, gap junction proteins, connexin-43, contraction-associated proteins, and stabilization of focal adhesions. The reinforcement of ligand–integrin interaction guarantees tight intercellular cohesion and electromechanical syncytial properties of the myometrium. Significant changes are also seen in the interstitial matrix with increased production of collagen type IV, fibronectin, and laminin $\beta2$. They grow into a regular fibrillar stroma and provide additional mechanical support to myocytes.

R.N. Miftahof and H.G. Nam, *Biomechanics of the Gravid Human Uterus*,
DOI 10.1007/978-3-642-21473-8_2, © Springer-Verlag Berlin Heidelberg 2011

As a result of the phenotypic modulation and fetal growth, at full term of gestation, the human uterus measures 40–42 cm in the axial (the fundal height) and 35–37 cm in the transverse directions. Ultrasonographic data show that the thickness of the wall in different regions varies between 0.45 and 0.7 cm and the radii of the curvature changes in the range of 8–14 cm (Celeste and Mercer 2008; Sfakiani et al. 2008; Degani et al. 1998). The fundal height reflects fetal growth and the increasing development of amniotic fluid in the amniotic sac – a two-layered membrane that surrounds the fetus. Amniotic fluid is primarily produced by the mother in the first trimester of pregnancy and after that by the fetus. It provides nourishment, allows free fetal movements inside the womb, and cushions the baby from external "blows". The amount of fluid is greatest at 34 weeks gestation, $\breve{V} \simeq 800$ ml and reduces to $\breve{V} \simeq 600$ ml in the following 6–7 weeks. Smaller or excessive amounts of amniotic fluid are called oligo- and polyhydramnious, respectively, and are usually manifestations of various medical conditions.

In the resting state, amniotic fluid generates intrauterine pressure, p, of 2–12 mmHg. During labor intrauterine pressure rises to 60–100 mmHg and is concurrent with contractile activity of the myometrium (Aelen 2005; Karash 1970; Seitchik and Chatkoff 1975). The pattern of changes in p serves as an important predictor of normal delivery.

The shape of the pregnant human womb is pear-like. Its actual configuration, though, before labor greatly depends on the position of the fetus and the amount of amniotic fluid. The most common presentation is vertex or head down, followed by oblique and transverse, when the baby lies obliquely and crosswise the uterus, respectively (Fig. 2.1). The head down position is normally delivered vaginally, while in the case of the oblique and transverse lie most babies are delivered by Cesarian section.

After parturition the uterus gradually returns almost to its usual size, but certain traces of its enlargement remain. Its cavity is larger than in the pre-pregnant state, vessels are tortuous, and muscular layers are more prominent.

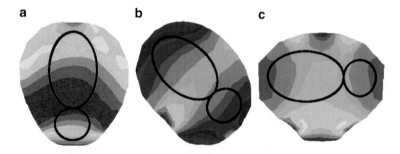

Fig. 2.1 Anatomical configurations of the pregnant uterus with fetus in (**a**) cephalic, (**b**) oblique, and (**c**) transverse presentation

2.2   Mechanical Models of the Gravid Uterus

In recent years, the biomedical research on reproduction has been focused mainly on molecular, neuroendocrine, and pharmacological aspects of uterine activity. A relatively small amount of work has been dedicated to modeling of the uterus per se. The knowledge of mechanical properties of the myometrium is crucial for the integration of motor functions into a biologically plausible biomechanical model of the organ. Experiments on quasistatic uniaxial stretching of linear human myometrial strips revealed nonlinear viscoelastic properties of the tissue. Thus, maximum deformability and tension recorded from the nonpregnant myometrium were $\varepsilon = 0.4$, $\sigma = 22\,\text{mN/mm}^2$, and at term, $\varepsilon = 0.8$, $\sigma = 41.6\,\text{mN/mm}^2$ (Wood 1964a, b; Conrad et al. 1966a, b; Pearsall and Roberts 1978; Buhimchi et al. 2007). Mechanical characteristics show a high degree of regional variability and hormonal dependency.

A large number of studies on isolated uterine tissue have been carried out on recording of electrically induced contractions. However, these results do not allow to reconstruct active, $T^a(\lambda)$, and passive, $T^p(\lambda)$, force–stretch ratio relationships which are of prime importance. Up to the time at which this book was being written, no experimental data were available on active uniaxial and biaxial stress–strain characteristics of the wall of the human uterus.[1] Hence, it is not surprising that biophysically credible models of the tissue as a mechanical continuum have not been constructed yet.

Under the assumptions of general isotropy and the zero stress–strain initial state, the constitutive equation for passive myometrium was derived by Mizrahi and Karni (1981)

$$
\sigma_s^r = \left[\left(-c_1 + \frac{c_3}{c_2} \right) - \frac{c_3}{c_2} \frac{1 + I - II}{1 + I - II + III} \right] \delta_s^r - \left[\frac{c_3}{c_2^2} \frac{1 + I}{1 + I - II + III} \right] \dot{\varepsilon}_s^r
$$
$$
+ \left[\frac{c_3}{c_2^3} \frac{1}{1 + I - II + III} \right] \dot{\varepsilon}_\alpha^r \, \dot{\varepsilon}_s^\alpha,
$$

(2.1)

where σ_s^r is the stress tensor, $\dot{\varepsilon}_s^r$ is the strain-rate tensor, δ_s^r is the Kronecker delta, I, II, and III are the principal invariants of the ratio $\dot{\varepsilon}_s^r / c_2$, and c_i ($i = \overline{1,3}$) are empirical constants. However, relevance and validity of the model were never examined. Suffice it to say that in their earlier study authors demonstrate that only during the first short stage of labor, the myometrium exhibits isotropic characteristics. In advanced labor and during delivery it behaved as a material with properties of curvilinear anisotropy (Mizrahi et al. 1978).

Various approaches have been developed for uterine mechanics modeling during parturition. The majority of them rely on assumptions of geometrical and physical

---

[1]Authors are familiar only with a single publication related to the study of dynamic biaxial properties of pregnant porcine uterine tissue by Manoogian et al. (2008).

linearity. In phenomenological models proposed by Vauge et al. (2000, 2003), the "uterus" was composed of N identical cells. At any given time, t, they are either in an active, $N_1(t)$, refractory, $N_2(t)$, or quiescent, $N_3(t)$, state. The system of differential equations was obtained

$$\frac{dN_1(t)}{dt} = \alpha N_3(t) - \frac{N_1(t)}{\tau}$$

$$\frac{dN_2(t)}{dt} = \frac{N_1(t)}{\tau} - \frac{N_1(t - t_r)}{\tau}$$ \hfill (2.2)

$$\frac{dN_3(t)}{dt} = \frac{N_1(t - t_r)}{\tau} - \alpha N_3(t), t \geq 0$$

$$0 \leq N_i(t), N_i(t) \leq N, \quad \text{where } i = 1, 2, 3$$

$$N = N_1(t) + N_2(t) + N_3(t).$$

Here, α is the excitability of the myocytes, τ is the natural lifetime of the active state, and t_r is the duration of the refractory state.

Results of numerical simulations obtained for different values of parameters and constants resembled patterns of spontaneous contractility observed during normal labor. For large t, $t \gg 1/\alpha, \tau$ and t_r, the population of cells attains constant values

$$n_1(\text{lim}) = \frac{\alpha \tau N}{1 + \alpha(\tau + t_r)}, \quad n_2(\text{lim}) = \frac{\alpha t_r N}{1 + \alpha(\tau + t_r)}, \quad n_3(\text{lim}) = \frac{N}{1 + \alpha(\tau + t_r)},$$
\hfill (2.3)

the amplitude of oscillations decreases, and the "uterus" succumbs to tonic contraction.

Bursztyn et al. (2007) simulated excitation contraction coupling in a single uterine smooth muscle cell. The calcium current accounted for the activity of voltage-gated Ca^{2+} channels

$$I_{Ca} = \frac{g_{Ca}(V^s - V_{Ca})}{1 + \exp(V_{Ca,1/2} - V/S_{Ca,1/2})},$$ \hfill (2.4)

where I_{Ca} is the Ca^{2+} current, V^s is the transmembrane potential, $V_{Ca,1/2}$ is the half-activation potential, $S_{Ca,1/2}$ is the slope factor for the Ca^{2+} current, and g_{Ca} is the maximal conductance of the channel. The net $\left[Ca_i^{2+}\right]$ was defined by

$$\frac{d\left[Ca_i^{2+}\right]}{dt} = I_{Ca} - I_{Ca,pump} + I_{Na/Ca},$$ \hfill (2.5)

where the efflux of Ca^{2+} through pumps was given by

$$I_{Ca,pump} = V_{p,max} \frac{[Ca_i^{2+}]^{n_M}}{[Ca_i^{2+}]_{1/2}^{n_M} + [Ca_i^{2+}]^{n_M}}, \tag{2.6}$$

and the Na^+/Ca^{2+} exchanger as

$$I_{Na/Ca} = g_{Na/Ca} \frac{[Ca_i^{2+}]}{[Ca_i^{2+}]_{Na/Ca} + [Ca_i^{2+}]} (V^s - V_{Na/Ca}). \tag{2.7}$$

In the above $[Ca_i^{2+}]_{1/2}^{n_M}$ and $[Ca_i^{2+}]_{Na/Ca}$ are concentrations of Ca^{2+} for half-activation of the pump and the exchanger, respectively, $V_{Na/Ca}$ is the reversal potential of the Na^+/Ca^{2+} exchanger, $V_{p,max}$ is the maximal velocity of ion extraction by the Ca^{2+} pump, $g_{Na/Ca}$ is the conductance, and n_M is the Hill coefficient.

The four-state cross-bridge model relied on the dynamics of myosin light chain phosphorylation and was described by (Hai and Murphy 1988)

$$\begin{aligned}
\frac{dF_M}{dt} &= -k_1(t)F_M + k_2 F_{Mp} + k_7 F_{AM} \\
\frac{dF_{Mp}}{dt} &= k_4 F_{AMp} + k_1(t)F_M - (k_2 + k_3)F_{Mp} \\
\frac{dF_{AMp}}{dt} &= k_3 F_{Mp} + k_6(t)F_{AM} - (k_4 + k_5)F_{AMp} \\
\frac{dF_{AM}}{dt} &= k_5 F_{AMp} - (k_7 + k_6(t))F_{AM},
\end{aligned} \tag{2.8}$$

with the constraint

$$F_M(t) + F_{Mp}(t) + F_{AMp}(t) + F_{AM}(t) = 1. \tag{2.9}$$

The stress, σ, in the myocyte was calculated as

$$\sigma = \sigma_{max}(F_{AM}(t) + F_{AMp}(t)). \tag{2.10}$$

Here, F_M, F_{Mp}, F_{AMp}, and F_{AM} are the fractional amounts of free unphosphorylated, phosphorylated (subscript p), attached phosphorylated, and attached dephosphorylated cross-bridges, respectively, and k_{1-7} are the rates of chemical reactions.

The excitation–contraction coupling is ensured by the dependence of the rate parameter of MLCK phosphorylation $k_1(t)$ and $[Ca_i^{2+}]$

$$k_1(t) = \frac{[Ca_i^{2+}]^{n_M}}{[Ca_{MLCK}^{2+}]_{1/2}^{n_M} + [Ca_i^{2+}]^{n_M}}, \tag{2.11}$$

where $\left[Ca^{2+}_{MLCK}\right]^{n_M}_{1/2}$ is the concentration of intracellular calcium required for half-activation of MLCK by Ca^{2+}-calmodulin.

Model simulations reproduced voltage-clamp traces recorded experimentally on pregnant rats and human nonpregnant myometrial cells.

In their pioneer work, Mizrahi and Karni (1975, 1978) modeled the gravid uterus as a thin axisymmetric elastic shell. The myometrium was treated as a homogenous isotropic biomaterial with general characteristics of smooth muscle tissue. The uterus was subjected to inner amniotic fluid pressure and its internal volume remained constant throughout. Deformations and displacements of the shell were small and only a single contraction was evaluated. The following kinematic relations were obtained:

$$\varepsilon^z_{11} = \frac{1}{(1+G^1_{31}x^3)}\left[\varepsilon_{11} + x^3\left(\frac{\partial}{\partial\alpha^1}\left(2\frac{g_{12}\varepsilon_{23}-\varepsilon_{13}}{H^2}\right)+\frac{\partial}{\partial\alpha^1}\left(2\frac{g_{12}\varepsilon_{13}-\varepsilon_{23}}{H^2}\right)g_{12}\right.$$
$$\left. +2\frac{g_{12}\varepsilon_{13}-\varepsilon_{23}}{H^2}G_{211}\right]$$

$$\varepsilon^z_{22} = \frac{1}{(1+G^2_{32}x^3)}\left[\varepsilon_{22} + x^3\left(\frac{\partial}{\partial\alpha^1}\left(2\frac{g_{12}\varepsilon_{13}-\varepsilon_{23}}{H^2}\right)+\frac{\partial}{\partial\alpha^2}\left(2\frac{g_{12}\varepsilon_{23}-\varepsilon_{13}}{H^2}\right)g_{12}\right.$$
$$\left. +2\frac{g_{12}\varepsilon_{23}-\varepsilon_{13}}{H^2}G_{122}\right]$$

$$2\varepsilon^z_{12} = \frac{1}{(1+G^2_{32}x^3)}\left[\omega_2 + x^3\left(\frac{\partial}{\partial\alpha^2}\left(2\frac{g_{12}\varepsilon_{23}-\varepsilon_{13}}{H^2}\right)+\frac{\partial}{\partial\alpha^2}\left(2\frac{g_{12}\varepsilon_{13}-\varepsilon_{23}}{H^2}\right)g_{12}\right.$$
$$\left. +2\frac{g_{12}\varepsilon_{13}-\varepsilon_{23}}{H^2}G_{212}\right] + \frac{1}{(1+G^1_{31}x^3)}\left[\omega_1 + x^3\left(\frac{\partial}{\partial\alpha^1}\left(2\frac{g_{12}\varepsilon_{13}-\varepsilon_{23}}{H^2}\right)\right.\right.$$
$$\left.\left. +\frac{\partial}{\partial\alpha^1}\left(2\frac{g_{12}\varepsilon_{23}-\varepsilon_{13}}{H^2}\right)g_{12}+2\frac{g_{12}\varepsilon_{23}-\varepsilon_{13}}{H^2}G_{121}\right],$$

$$(2.12)$$

where (α^1, α^2) are the coordinate lines on the middle surface (S) of the shell, x^3 is the normal coordinate to S, g_{12} is the mixed component of the metric tensor, ε_{ij} and ε^z_{ij} $(i,j = 1,2)$ are the strain tensors defined on the middle and equidistant, S_z, surfaces, respectively, $\omega_i (i = 1, 2)$ is the rotation vector, G^i_{jk} is the geodesic torsion, and G_{iji} and G^j_{ii} $(i,j,k = 1,2,3)$ are the normal and geodesic curvatures, respectively.

The initial strain values at four distinct points on S_z were defined from measurements recorded from the abdominal surface of pregnant women in labor. The system (2.12) was solved numerically. Time variations of the radii of curvature and torsions along the longitudinal and circular fibers were assessed. Results revealed that during the most intense contraction the curvature in the longitudinal direction increased by 200%, and in the circumferential direction only by 20%. These changes were associated with the transformation of the uterus from a pear-like to more spherical shape.

Later, İrfanoğlu and Karaesmen (1993) modeled the uterus as a thin ellipsoid of revolution. The myometrium was treated as a homogenous, isotropic, and incompressible material. Deformations and displacements were finite. The equilibrium equations of the shell in terms of the undeformed reference configuration $(\bar{r}_1, \bar{r}_2, \phi)$ were given by (Flügger and Chou 1967)

$$\frac{d\psi}{d\phi} = \frac{P_n r_1 \lambda_\theta \lambda_\phi}{N_\theta} - \frac{N_\theta r_1}{N_\phi r} \sin\psi$$

$$\frac{d\lambda_\phi}{d\phi} = \left(\frac{dN_\phi}{d\phi}\right)^{-1} \left(N_\theta \cos\phi - N_\phi \frac{r_1}{r} \cos\psi\right) - \frac{dN_\phi}{d\theta} \left(\frac{dN_\phi}{d\phi}\right)^{-1} \frac{d\lambda_\phi}{d\phi} \qquad (2.13)$$

$$\frac{d\lambda_\theta}{d\phi} = \frac{r_1}{r} \left(\lambda_\phi \cos\psi - \lambda_\theta \cos\phi\right),$$

where P_n is the internal amniotic fluid pressure, $\bar{r}_1, \bar{r}_2, \phi$ and ψ are geometric parameters, λ_ϕ and λ_θ are the stretch ratios, and N_ϕ and N_θ are the in-plane forces per unit length in the longitudinal and circumferential directions, respectively. Two types of constitutive relationships were considered, i.e., Mooney–Rivlin

$$N_\phi = \frac{2}{\lambda_\phi} \left(\lambda_\phi^2 - \lambda_\theta^{-2}\lambda_\phi^{-2}\right)\left(1 + \underline{\alpha}\lambda_\theta^2\right)$$

$$N_\theta = \frac{2}{\lambda_\theta} \left(\lambda_\theta^2 - \lambda_\theta^{-2}\lambda_\phi^{-2}\right)\left(1 + \underline{\alpha}\lambda_\phi^2\right), \underline{\alpha} - const \qquad (2.14)$$

and neo-Hookean $(\underline{\alpha} = 0)$

$$N_\phi = \frac{2}{\lambda_\phi} \left(\lambda_\phi^2 - \lambda_\theta^{-2}\lambda_\phi^{-2}\right)$$

$$N_\theta = \frac{2}{\lambda_\theta} \left(\lambda_\theta^2 - \lambda_\theta^{-2}\lambda_\phi^{-2}\right). \qquad (2.15)$$

The lower segment and the apex of the uterus were clamped and the organ maintained axial symmetry throughout. The effect of physical nonlinearities and the initial configuration on intrauterine pressure and volume were analyzed.

Weiss et al. (2004) developed a three-dimensional finite element-based model of the uterus. The stress–strain distribution in the organ under static conditions with regard to internal amniotic pressure was investigated. An agreement between theoretical results and in vivo clinical measurements was achieved through multiple numerical experiments and adjustments of input parametric data.

Rice et al. (1975, 1976) studied the process of cervix opening during the first short stage of labor. The cervix was modeled as an axisymmetric orthotropic thin membrane. The tissue was assumed to be homogenous, incompressible biocomposite and possessed exponentially elastic and viscoelastic properties. Deformations were small, but displacements were finite. The head was in the vertex position and exerted dilatational constant pressure on the membrane. No friction condition was imposed between the head and the cervix. The axial load was defined a priori as a function of time.

The four degrees-of-freedom linear element approximation was solved for deflection confined by sliding curved restraints. The model results predicted an elastic stretch and a small viscous strain during a single contraction, a monotonic relationship between internal pressure and dilatation, and continuous opening of the cervix following unimpeded fetal descent. However, no comparison to real experimental data was made.

Paskaleva (2007) investigated experimentally and studied theoretically the biomechanics of cervical insufficiency – a medically worrisome condition related to a dilatation of the cervix in absence of myometrial contractions. The cervix constituted a viscoelastic two-compartment continuum. The first compartment consisted of connective tissue, i.e., collagen and elastin fibers, embedded into the hydrated ground substance of glycosaminoglycans and proteoglycans. The second was represented by interstitial fluid. Diffusion of interstitial fluid between the two compartments satisfied Darcy's Law. The constitutive model accounted mechanics of individual components and captured tissue growth and remodeling. It had the form

$$\widetilde{\sigma} = \frac{1}{J}\left(f_1^0 J_1 \sigma_1 + f_2^0 J_2 \sigma_2\right) + \sigma_{\text{el}} + \Delta p \tag{2.16}$$

where

$$\sigma_1 = \sigma_c + \sigma_{\text{BG}} + \sigma_{\text{IC}}, \quad \sigma_2 = \sigma_{\text{FG}}.$$

Here, $\widetilde{\sigma}$ is the macroscopic Cauchy stress in the cervical stroma, f_i^0 is the volume fractions, $\widetilde{J}$ and J_i are the total and compartmental volumetric stretches ($i = 1, 2$ indicates compartments), and p is the hydrostatic pressure. Macroscopic stresses in the collagen and elastin networks, σ_c, σ_{el}, stresses produced by bound, σ_{BG}, and free, σ_{FG}, aminoglycans, and inter-compartmental pressure, σ_{IC}, were given by

$$\begin{aligned}
\sigma_c &= \frac{\mu \lambda_L}{\xi^3 J_1}\left[\frac{1}{\lambda_c}\beta\left(\frac{\lambda_L}{\lambda_c}\right)\mathbf{B}_c - \beta_0 \mathbf{I}\right] \\
\sigma_{\text{el}} &= B_{\text{el}}(\widetilde{J} - 1)\mathbf{I} \\
\sigma_j &= B_{\text{GAG}} \ln(J_j) \quad (j = \text{BG,FG}) \\
\sigma_{IC} &= \sigma_0 \sqrt[m]{\dot{\varepsilon}_v / \dot{\varepsilon}_v^0}\,.
\end{aligned} \tag{2.17}$$

Here, $\mathbf{B}_c$ and $\mathbf{I}$ are the left Cauchy–Green and identity tensors, respectively, of the gradient of deformation, μ is the initial collagen modulus, λ_L and λ_c are the maximal (L) and current (c) stretch ratios of a collagen fibril, B_{el} and B_{GAG} are the bulk moduli of elastin and glycosaminoglycans, σ_0 is the flow strength, $\dot{\varepsilon}_v^0$ and $\dot{\varepsilon}_v$

are the initial and current volumetric flow rates, and ξ, m, β, and β_0 are structural parameters.

The configuration of the uterus and surrounding anatomical structures were reconstructed digitally from a set of magnetic resonance images obtained from gravid women. It was assumed that the uterus underwent axisymmetric deformations throughout. Contractions of the myometrium were mimicked by varying the amniotic pressure vs. time. Stress–strain distribution in the organ was analyzed using a commercial finite element solver. Continuum four-node tetrahedron elements were adopted to model the uterus and three-node triangular general-purpose shell elements were chosen to simulate the amniotic membrane. Mechanical parameters were estimated from in vitro experiments on samples excised from pregnant human uteri. The "missing" data were amended during numerical simulations. Results provided insight into the stress–strain distribution in the cervical region and the "dynamics" of dilatation at quasistatic states.

2.3 Models of Myoelectrical Activity

A plausible model of electrical activity of a myocyte was developed by Rihana et al. (2006, 2009). It employed general principles of the Hodgkin–Huxley formalism

$$C_m \frac{dV^s}{dt} = -(I_{Na} + I_{Ca} + I_{K_v} + I_{Ca-K} + I_{Cl}) + I_{stim}. \tag{2.18}$$

Here, $I_{Na}, I_{Ca}, I_{K_v}, I_{Ca-K}$, and I_{Cl} are the voltage-gated Na$^+$, Ca^{2+}, K$_v$ (a mix of three different types of potassium), Ca^{2+}-activated K$^+$, and Cl$^-$ currents, respectively, I_{stim} is the stimulus current, and C_m is the specific membrane capacitance. Each current was related to the membrane voltage, reversal potentials, V_i, for $i = $ Na$^+$, Ca^{2+}, K$^+$, and Cl$^-$ ions, the specific conductance, g_i, and gating variables m_i, n_i, and h_i as

$$
\begin{aligned}
I_{Na} &= g_{Na} m_{Na}^2 h_{Na} (V^s - V_{Na}) \\
I_{Ca-K} &= g_{K(Ca)} \left(\frac{[Ca_i^{2+}]^n}{[Ca_i^{2+}]_{1/2}^n + [Ca_i^{2+}]^n} \right) (V^s - V_{Na}) \\
I_{Ca} &= g_{Ca} m_{Ca}^2 h_{1Ca} h_{2Ca} (V^s - V_{Ca}) \\
I_K &= g_K n_{K1} n_{K2} h_{K1} (V^s - V_K) \\
I_{Cl} &= g_{Cl} (V^s - V_{Cl}).
\end{aligned}
\tag{2.19}
$$

The dynamics of intracellular calcium was described by

$$\frac{d[Ca_i^{2+}]}{dt} = f_c(\alpha I_{Ca} - K_{Ca}[Ca_i^{2+}]). \tag{2.20}$$

Here, f_c is the probability of the influx of Ca^{2+}, $\hat{\alpha}$ is the conversion factor, and K_{Ca} represents the sequestration, extrusion, and buffering processes of calcium by intracellular compartments.

The variation of m_i, n_i, and h_i satisfied the first-order differential equations

$$
\begin{aligned}
\frac{dm_{Na}}{dt} &= \frac{1}{\tau_{mNa}}(m_{Na\infty} - m_{Na}), & \frac{dh_{Na}}{dt} &= \frac{1}{\tau_{hNa}}(h_{Na\infty} - h_{Na}), \\
\frac{dm_{Ca}}{dt} &= \frac{1}{\tau_{mCa}}(m_{Ca\infty} - m_{Ca}), & \frac{dh_{1Ca}}{dt} &= \frac{1}{\tau_{h1Ca}}(h_{1Ca\infty} - h_{1Ca}), \\
\frac{dh_{2Ca}}{dt} &= \frac{1}{\tau_{h1Ca}}(h_{2Ca\infty} - h_{2Ca}), & \frac{dn_{K1}}{dt} &= \frac{1}{\tau_{K1}}(n_{K1\infty} - n_{K1}), \\
\frac{dn_{K2}}{dt} &= \frac{1}{\tau_{K2}}(n_{K2\infty} - n_{K2}), & \frac{dh_{K1}}{dt} &= \frac{1}{\tau_{h1}}(h_{K1\infty} - h_{K1}),
\end{aligned}
\tag{2.21}
$$

where

$$
\begin{aligned}
m_{Na\infty} &= 1 \left/ \left(1 + \exp\frac{(V^s + 21)}{-5}\right)\right., & h_{Na\infty} &= 1 \left/ \left(1 + \exp\frac{(V^s + 58.9)}{8.7}\right)\right., \\
m_{Ca\infty} &= 1 \left/ \left(1 + \exp\frac{(V^s + V_{Ca,1/2})}{S_{Ca}}\right)\right., & h_{1Ca\infty} &= 1 \left/ \left(1 + \exp\frac{(V^s + 34)}{5.4}\right)\right., \\
n_{Kj\infty} &= 1 \left/ \left(1 + \exp\frac{(V^s + V_{Kj,1/2})}{S_{Kj}}\right)\right., & h_{K1\infty} &= 1 \left/ \left(1 + \exp\frac{(V^s + V_{hK1,1/2})}{S_{hK1}}\right)\right..
\end{aligned}
\tag{2.22}
$$

Here, $V_{Ca,1/2}, V_{Kj,1/2}$, and $V_{hK1,1/2}$ are the half-activation potentials, and S_{Ca}, S_{Kj}, and S_{hK1} $(j = 1, 2)$ are the slope factors for specific currents. Time constants

$$
\begin{aligned}
\tau_{mNa} &= 0.25\exp(-0.02V^s) \\
\tau_{hNa} &= 0.22\exp(-0.06V^s) + 0.366 \\
\tau_{mCa} &= 0.64\exp(-0.04V^s) + 1.188 \\
\tau_{h1Ca}, \tau_{Kj}, \tau_{h1} &= \text{const},
\end{aligned}
\tag{2.23}
$$

were adjusted during simulations.

The model reproduced a variety of electrical patterns, i.e., slow wave and bursting with action potential generation, recorded in myometrium at term.

In their later model of the propagation of the wave of excitation, $V^s = V^s(x, y)$, in a two-dimensional electrically isotropic uterine syncytium, only three ion currents were retained (Rihana et al. 2007)

$$
C_m\frac{\partial V^s}{\partial t} = \frac{1}{R_a}\nabla V^s - (I_{Ca} + I_{K_v} + I_{Ca-K}) + I_{stim},
\tag{2.24}
$$

Here, $\nabla (\nabla = \frac{\partial}{\partial x}\bar{i} + \frac{\partial}{\partial y}\bar{j})$ is the spatial gradient operator, R_a is the axial syncytial resistance, and the meaning of other parameters as described above. The dynamics of I_{Ca}, I_{K_v}, and I_{Ca-K} currents were given by (2.19)–(2.22).

Results of numerical simulations revealed the preferred axial spread of excitation. The amplitude of V decreased away from the pacemaker. The predicted conduction velocity corresponded to the value obtained experimentally from the rat myometrium at term.

La Rosa et al. (2009) proposed a bidomain model of the abdomen with an enclosed uterus to study the propagation of the wave of depolarization, $V^s = V^s V(\bar{r}, t)$, in myometrium and the induced magnetic field, $\overrightarrow{B}$, on the abdominal surface

$$C_m \frac{\partial V^s}{\partial t} = \left(G_i - \frac{G_i}{G_i + G_e}\right)\nabla^2 V^s - I_{ion} + I_{stim}$$
$$\nabla \times \overrightarrow{B} = -G_i(\mu_A + \sigma_A)\nabla V^s. \tag{2.25}$$

Here, G_i and G_e are the intracellular and extracellular conductivities and μ_A and σ_A are the permeability and conductivity constants of the abdominal space. The uterus was considered as an electrically anisotropic sphere of a radius $\bar{r}$. The pacemaker cell was located in the polar region. A modified form of the Fitzhugh–Nagumo equations was employed to describe the ion current dynamics (FitzHugh 1961; Nagumo et al. 1962)

$$I_{ion} = k(V^s - V_a)(V^s - 1)V^s - v^*$$
$$\frac{dv^*}{dt} = \varepsilon(V^s - \gamma v^*), \tag{2.26}$$

where v^* is the recovery variable and k, V_a, ε, and γ are empirical parameters.

The model supposedly reproduced electromyograms and magnetograms recorded from the pregnant uterus. However, the results of numerical simulations were inconclusive.

Investigations into the conditions and morphostructural principles for autorhythmicity proved that the organization of the myometrium is of fundamental importance in the generation, maintenance, and propagation of the wave of excitation within it. Multiple components of interconnected signaling pathways play various roles in the temporal features of the oscillation phenomenon. By assuming a linear coupling by gap junctions, their evolution was governed by (Jacquemet 2006; Joyner et al. 2006)

$$C_{mi} \frac{dV_i^s}{dt} = -I_{ion,i} + (-1)^i \tilde{g}(V_1 - V_2) \quad i = 1, 2 \tag{2.27}$$

and clearly demonstrated a possibility of bursting activity. In the above, $i = 1$ is referred to a group of excitable, $i = 2$ – nonexcitable cells, and $\tilde{g}$ is the coupling conductance. The ion current $I_{ion,1}$ was defined by (2.26), and $I_{ion,2} = g_m(V_2 - V_r)$, where g_m is the membrane conductance and V_r is the resting membrane potential. Sensitivity analysis revealed that ion channels with fast activation/deactivation dynamics were responsible for spike production. Essential elements for the existence and sustainability of pacemaker activity were relative uncoupling of the pacemaker from the surrounding tissue, the presence of a gradual transition zone, and distributed tissue anisotropy.

Model analysis of synchronization and bursting in the uterine muscle in late pregnancy revealed four types of spatial automaticity: (1) sparse, with reduced amplitude of spikes; (2) clustered, when a group of cells acted as an independent local pacemaker; (3) uniform, with a constant depolarization or small amplitude spike oscillations; and (4) coherent, with a synchronous discharge of spatially distributed pacemakers (Benson et al. 2006). The specific pattern depended entirely on the strength of cellular coupling.

The existing models of the myometrium and the gravid human uterus, as described above, are based on the application of general principles of solid mechanics and the theory of thin elastic shells. They incorporate some morphological data on the structure and function of the organ and thus hold a promise for the area of biomechanics of the uterus during labor and delivery. Although these models are of limited biomedical value, they serve as a platform for further development of integrative, biologically plausible models.

Exercises

1. Biological tissues and organs can experience stresses in the absence of deformations and vice versa. This makes it difficult to define their initial configuration. However, the gravid human uterus close to term (~34 weeks) can be considered stress and deformation free. What anatomical and morphological changes support the assumption? (*Hint*: recall phases of "phenotypic modulation" of the uterus).
2. The uterus undergoes significant geometrical changes throughout pregnancy. At what stage does it start enduring deformations actually? (*Hint*: recall phases of "phenotypic modulation" of the uterus).
3. Why is the modeling of the pregnant human uterus so "difficult"?
4. The Hogdkin–Huxley and Fitzhugh–Nagumo models are widely used to study electrical processes in nerve axons, excitable cells, and tissues. What are biological and mathematical pros and cons of each approach?
5. Study physicochemical changes in the fibrillar stroma and extracellular matrix of the human cervix during "ripening". How do these changes affect mechanical properties of the tissue? (Åkerud 2009)
6. The Belousov–Zhabotinsky reaction is an example of self-excitatory and sustained oscillatory activity in a chemically reactive medium. It has been shown that similar patterns also emerge in a cell-chemical system through self-organization. Does the system (2.18)–(2.23) have periodic solutions?
7. Can the uterus in labor have multiple pacemaker zones?

8. Uterine electromyography and magnetography are noninvasive techniques that record electromagnetic myometrial activity. Traces have a diagnostic value and help decide on rational therapeutic management. The inverse problem in electromyography and magnetography is the reconstruction of underlying cellular and subcellular mechanisms. Could that be done?
9. By what principle could one possibly judge a model of the gravid human uterus as biologically plausible and clinically valuable?

Chapter 3
A Dynamic Model of the Fasciculus

The further a mathematical theory is developed, the more harmoniously and uniformly does its construction proceed, and unsuspected relations are disclosed between hitherto separated branches of the science.

D. Hilbert

3.1 Formulation of the Model

To fully appreciate the biological phenomena that underlie processes of electrome-chanical activity in the pregnant uterus, we begin with analysis of basic myoelectrical events. We proceed from a one-dimensional numerical simulation of the dynamics of fasciculus (myofiber) – the functional unit of the human uterus.

Let a fasciculus be embedded in the extracellular matrix of connective tissue. Our developments of a biomechanical model will be based on the following assumptions which are consistent with several lines of experimental evidence (see Chap. 1).

1. Smooth muscle cells in the fasciculus are connected by tight junctions to form a homogenous electromechanical biological continuum; it is treated as a soft fiber thread.
2. The myofiber possesses nonlinear viscoelastic properties; the mechanics of inactive smooth muscle cells, collagen, elastin fibers, and ECM define the "passive," $T^p(\lambda, c_i)$, and intracellular contractile proteins describe the "active" component, $T^a(\lambda, Z_{mn}^{(*)}, [Ca_i^{2+}], c_i)$, of the total force, T^t

$$T^t = T^p(\lambda, c_i) + T^a(\lambda, Z_{mn}^{(*)}, [Ca_i^{2+}], c_i), \tag{3.1}$$

where λ is the stretch ratio, c_i are empirical material constants, $Z_{mn}^{(*)}$ is the "biofactor," and $[Ca_i^{2+}]$ is the concentration of free cytosolic calcium.
3. Contractions of the myofiber are isometric and deformations are finite.

R.N. Miftahof and H.G. Nam, *Biomechanics of the Gravid Human Uterus*, DOI 10.1007/978-3-642-21473-8_3, © Springer-Verlag Berlin Heidelberg 2011

4. Myogenic electrical events are a result of activity of an intrinsic autonomous oscillator; its function is defined by the fast (T-type) and slow (L-type) inward Ca^{2+}, BK_{Ca}, voltage-dependent K_{v1}^{+} and leak Cl^{-} currents.
5. Each oscillator is in the silent state; the transformation to a firing state is a result of depolarization and/or stretch deformation of the cell that alters the conductance for L- and T-type Ca^{2+} channels, while the stretch affects permeability of L-type channels.
6. The myofiber possesses cable electrical properties; propagation of the wave of depolarization is a result of combined activity of the Na^{+}, K_{v2}^{+}, and leak Cl^{-} ion currents.
7. A smooth muscle cell or a group of cells within the fasciculus have intrinsic pacemaker properties; the transformation from a silent to a bursting state can occur spontaneously and is a result of "alterations" in electrical properties; additionally, an a priori defined "pacemaker" provides an excitation to the fiber.

Let the fasciculus of a length L be referred to a local Lagrange coordinate system α. Its equation of motion is given by:

$$\rho \frac{\partial v}{\partial t} = \frac{\partial}{\partial \alpha} T^{t}, \quad (0 \leq \alpha \leq L) \tag{3.2}$$

where ρ is density, v is the velocity, and the meaning of other parameters are as described above. Following the working assumption (2), the total force T^{t} can be decomposed as:

$$T^{t} = k_{v} \frac{\partial(\lambda - 1)}{\partial t} + T^{a}(\lambda, Z_{mn}^{(*)}, [Ca_{i}^{2+}], c_{i}) + T^{p}(\lambda, c_{i}), \tag{3.3}$$

where the viscoelastic term has been added to (3.1). Here k_{v} is viscosity. Substituting the above into (3.3) we obtain

$$\rho \frac{\partial v}{\partial t} = \frac{\partial}{\partial \alpha} \left(k_{v} \frac{\partial(\lambda - 1)}{\partial t} + T^{a}(\lambda, Z_{mn}^{(*)}, [Ca_{i}^{2+}], c_{i}) + T^{p}(\lambda, c_{i}) \right), \tag{3.4}$$

where the force–stretch ratio relationship yields

$$T^{p} = \begin{cases} c_{1}[\exp c_{2}(\lambda - 1) - 1], & \lambda > 1.0, \\ 0, & \text{otherwise.} \end{cases} \tag{3.5}$$

and the active force–intracellular Ca_{i}^{2+} relationship for the myometrium is given by:

$$T^{a} = \begin{cases} 0, & [Ca_{i}^{2+}] \leq 0.1 \, \mu M \\ c_{3} + c_{4}[Ca_{i}^{2+}]^{4} + c_{5}[Ca_{i}^{2+}]^{3} + c_{6}[Ca_{i}^{2+}]^{2} + c_{7}[Ca_{i}^{2+}], & \\ & 0.1 < [Ca_{i}^{2+}] \leq 1 \, \mu M \\ \max T^{a}, & [Ca_{i}^{2+}] > 1 \, \mu M. \end{cases} \tag{3.6}$$

The system of equations for the oscillatory activity of the membrane potential V is

$$\hat{\lambda}\, C_{\mathrm{m}} \frac{\mathrm{d}V}{\mathrm{d}t} = -\sum_j \tilde{I}_j \tag{3.7}$$

where $\hat{\lambda}$ is the numerical parameter, C_{m} is the uterine smooth muscle cell membrane capacitance, and $\tilde{I}_j$ is the sum of the respective ion currents.

$$
\begin{aligned}
\tilde{I}_{\mathrm{Ca}}^f &= g_{\mathrm{Ca}}^f \tilde{m}_i^3 \tilde{h}(V - V_{\mathrm{Ca}}), \\
\tilde{I}_{\mathrm{Ca}}^s &= g_{\mathrm{Ca}}^s \tilde{x}_{\mathrm{Ca}}(V - V_{\mathrm{Ca}}), \\
\tilde{I}_{\mathrm{K1}} &= g_{\mathrm{K1}} \tilde{n}^4 (V - V_{\mathrm{K1}}), \\
\tilde{I}_{\mathrm{Ca-K}} &= \frac{g_{\mathrm{Ca-K}}^f \left[\mathrm{Ca}_i^{2+}\right](V - V_{\mathrm{Ca}})}{0.5 + \left[\mathrm{Ca}_i^{2+}\right]}, \\
\tilde{I}_{\mathrm{Cl}} &= g_{\mathrm{Cl}}(V - V_{\mathrm{Cl}}).
\end{aligned}
\tag{3.8}
$$

Here $V_{\mathrm{Ca}}, V_{\mathrm{K1}}, V_{\mathrm{Cl}}$ are the reversal potentials, and $g_{\mathrm{Ca}}^f, g_{\mathrm{Ca}}^s, g_{\mathrm{K1}}, g_{\mathrm{Ca-K}}, g_{\mathrm{Cl}}$ are the maximal conductances for the ion currents, $\tilde{m}, \tilde{h}, \tilde{n}$ and $\tilde{x}_{\mathrm{Ca}}$ are dynamic variables described by

$$
\begin{aligned}
\tilde{m}_1 &= \frac{\tilde{\alpha}_m}{\tilde{\alpha}_m + \tilde{\beta}_m}, \\
\hat{\lambda}\, \tilde{h} \frac{\mathrm{d}\tilde{h}}{\mathrm{d}t} &= \tilde{\alpha}_h\left(1 - \tilde{h}\right) - \tilde{\beta}_h \tilde{h}, \\
\hat{\lambda}\, \tilde{h} \frac{\mathrm{d}\tilde{n}}{\mathrm{d}t} &= \tilde{\alpha}_n(1 - \tilde{n}) - \tilde{\beta}_n \tilde{n}, \\
\hat{\lambda}\, \tau_{x\mathrm{Ca}} \frac{\mathrm{d}\tilde{x}_{\mathrm{Ca}}}{\mathrm{d}t} &= \frac{1}{\exp(-0.15(V + 50))} - \tilde{x}_{\mathrm{Ca}}, \\
\hat{\lambda}\, \frac{\mathrm{d}\left[\mathrm{Ca}_i^{2+}\right]}{\mathrm{d}t} &= \wp_{\mathrm{Ca}} \tilde{x}_{\mathrm{Ca}}(V_{\mathrm{Ca}} - V) - \left[\mathrm{Ca}_i^{2+}\right]
\end{aligned}
\tag{3.9}
$$

where the activation $\tilde{\alpha}_y$ and deactivation $\tilde{\beta}_y$ $(y = \tilde{m}, \tilde{h}, \tilde{n})$ parameters of ion channels satisfy the empirical relations

$$
\begin{aligned}
\tilde{\alpha}_m &= \frac{0.1(50 - \tilde{V})}{\exp(5 - 0.1\tilde{V}) - 1}, & \tilde{\beta}_m &= 4\exp\frac{(25 - \tilde{V})}{18}, \\
\tilde{\alpha}_h &= 0.07\exp\frac{(25 - 0.1\tilde{V})}{20}, & \tilde{\beta}_h &= \frac{1}{1 + \exp(5.5 - 0.1\tilde{V})}, \\
\tilde{\alpha}_n &= \frac{0.01(55 - \tilde{V})}{\exp(5.5 - 0.1\tilde{V}) - 1}, & \tilde{\beta}_n &= 0.125\exp\frac{(45 - \tilde{V})}{80}.
\end{aligned}
\tag{3.10}
$$

Here $\tilde{V} = (127V + 8,265)/105$, τ_{xCa} is the time constant, $\wp_{Ca}$ is the parameter referring to the dynamics of Ca^{2+} channels, and $\hbar$ is a numerical constant.

The evolution of L- and T-type Ca^{2+}-channels depends on the wave of depolarization, V^s, and is defined by

$$g_{Ca}^s(t) = [\delta(V) + (\lambda(t) - 1)](\max g_{Ca}^s),$$
$$g_{Ca}^f(t) = (\lambda(t) - 1)g_{Ca}^f, \qquad (3.11)$$

where

$$\lambda(t) \geq 1.0, \quad \delta(V) = \begin{cases} 1, & \text{for} \quad V \geq V_p^s \\ 0, & \text{otherwise} \end{cases}.$$

Here V_p^s is the threshold value for V^s.

The propagation of the wave of excitation V^s is described by

$$C_m \frac{\partial V^s}{\partial t} = \frac{d_m}{R_s} \frac{\partial}{\partial \alpha} \left(\lambda(\alpha) \frac{\partial V^s}{\partial \alpha} \right) - (I_{Na} + I_{K2} + I_{Cl}), \qquad (3.12)$$

where d_m is the diameter, R_s is the specific resistance of the fasciculus, and

$$I_{Na} = g_{Na} \hat{m}^3 \hat{h} (V^s - V_{Na})$$
$$I_{K2} = g_{K2} \hat{n}^4 (V^s - V_{K2}) \qquad (3.13)$$
$$I_{Cl} = g_{Cl} (V^s - V_{Cl}).$$

Here g_{Na}, g_{K2}, g_{Cl} are the maximal conductances, and V_{Na}, V_{K2}, V_{Cl} are the reversal potentials of Na^+, K_{v2}^+, and Cl^- membrane currents, respectively. The dynamics of the variables $\hat{m}, \hat{h}, \hat{n}$ is described by:

$$\frac{d\hat{m}}{dt} = \hat{\alpha}_m(1 - \hat{m}) - \hat{\beta}_m \hat{m}$$
$$\frac{d\hat{h}}{dt} = \hat{\alpha}_h(1 - \hat{h}) - \hat{\beta}_h \hat{h} \qquad (3.14)$$
$$\frac{d\hat{n}}{dt} = \hat{\alpha}_n(1 - \hat{n}) - \hat{\beta}_n \hat{n}$$

with the activation $\hat{\alpha}_y$ and deactivation $\hat{\beta}_y$ $\left(y = \hat{m}, \hat{h}, \hat{n} \right)$ parameters given by

$$\hat{\alpha}_m = \frac{0.005(V^s - V_m)}{\exp 0.1(V^s - V_m) - 1}, \quad \hat{\beta}_m = 0.2 \exp \frac{(V^s + V_m)}{38},$$
$$\hat{\alpha}_h = 0.014 \exp \frac{-(V_h + V^s)}{20}, \quad \hat{\beta}_h = \frac{0.2}{1 + \exp 0.2(V_h - V^s)}, \qquad (3.15)$$
$$\hat{\alpha}_n = \frac{0.006(V^s - V_n)}{\exp 0.1(V^s - V_n) - 1}, \quad \hat{\beta}_n = 0.75 \exp(V_n - V^s).$$

Here, V_m, V_h, V_n are the reversal potentials for activation and inactivation of Na$^+$ and K$_{v2}^+$ ion currents of the myometrium.

In the following numerical experiments, we assume that at the initial moment of time the functional unit is in unexcitable state

$$V^s(\alpha, 0) = 0, v(\alpha, 0) = 0, \left[Ca_i^{2+}\right] = \overset{0}{\left[Ca_i^{2+}\right]}, \tag{3.16}$$
$$\hat{m} = \hat{m}_\infty, \hat{h} = \hat{h}_\infty, \hat{n} = \hat{n}_\infty, \tilde{h} = \tilde{h}_\infty, \tilde{n} = \tilde{n}_\infty, \tilde{x}_{Ca} = \tilde{x}_{Ca}^\infty.$$

It is activated by a series of discharges of action potentials by an intrinsic pacemaker cell

$$V^s(0, t) = \begin{cases} \overset{0}{V^s}, & 0 < t < t^d, \\ 0, & t \geq t^d \end{cases} \qquad V^s(0, t) = V(t). \tag{3.17}$$

The ends of the myofiber are clamped and remain unexcitable throughout

$$V^s(0, t) = V^s(L, t) = 0, v(0, t) = v(L, t) = 0. \tag{3.18}$$

Equations (3.4)–(3.15), complemented by initial and boundary conditions (3.16)–(3.18), constitute the mathematical formulation of the model of the electro-mechanical activity of the myometrial fasciculus. It describes:

1. Self-oscillatory behavior and/or myoelectrical activity induced by discharges of a "pacemaker" cell
2. Generation and propagation of the wave of depolarization along the myofiber
3. Coupling of spatially distributed oscillators
4. Generation of action potentials
5. Dynamics of the cytosolic Ca^{2+} transients
6. Active and passive force generation
7. Deformation of the fasciculus and the following excitation of the cell membrane with contractions

The governing system was solved numerically using ABS Technologies$^®$ computational platform. It employed a hybrid finite difference scheme and finite element method of second-order accuracy, with respect to spatial and time variables. The parameters and constants used in simulations were derived from the published literature. The values, which could not be found in the literature, were adjusted during the experiments in order to mimic closely the behavior of the biological prototype.

3.2   Effect of Changes in the Ionic Environment on Myoelectrical Activity

3.2.1   Physiological Condition

The resting membrane potential of the fasciculus is $V^r = -59$ mV. Continuous fluctuations at low rate and amplitudes of the L-type -0.08 nA, T-type Ca^{2+} -0.48 nA, respectively, an outward K^+ -0.03 nA, the BK_{Ca} -0.62 nA, and the small chloride -0.04 nA currents result in oscillations of the membrane potential V known as slow waves. Their frequency, $v = 0.02$ Hz, and the amplitude, $V = 27$ mV, remain constant. The maximum rate of depolarization is calculated as 9 mV/s and of repolarization $- 7.5$ mV/s.

The slow wave induces the flux of Ca^{2+} ions inside the cell at a rate of 0.057 μM/s. There is a 20-s time delay in the intracellular calcium transients as compared with the wave of depolarization. Free cytosolic calcium at $\max[Ca_i^{2+}] = 0.44$ μM activates the contractile protein system with the production of spontaneous contractions, $T^a = 13.6$ mN/cm (Fig. 3.1). They follow in phase and time the dynamics of calcium oscillations and are normally preceded by slow waves.

High-frequency discharges of an intrinsic pacemaker initiate high-magnitude ion currents: $\tilde{I}_{Ca}^s = 0.4, \tilde{I}_{Ca}^f = 0.51, \tilde{I}_{K1} = 0.2, \tilde{I}_{Ca-K} = 1.0$, and $\tilde{I}_{Cl} = 0.5$(nA) and the generation of action potentials of amplitudes 38/45 mV at a frequency of 2.7 Hz. A concomitant rise in intracellular calcium to 0.51 μM causes the development of active force, 16.6 mN/cm. Upon the termination of electrical discharges, the myofiber returns to its unexcited state.

3.2.2   Changes in Ca_0^{2+}

A gradual increase in extracellular calcium leads to depolarization of the membrane. Concentrations of Ca_0^{2+} 3–5 times normal cause the upshift of the resting potential to -34 mV and -31 mV (Fig. 3.2). This is associated with the exponential rise in free intracellular calcium to 0.72 μM. The myometrium undergoes tonic contraction, $T^a = 23.75$ mN/cm..

A concurrent electrical stimulation of the myofiber evokes ion currents of intensities: $\tilde{I}_{Ca}^s = 1.3, \tilde{I}_{Ca}^f = 0.61; \tilde{I}_{K1} = 0.62, \tilde{I}_{Ca-K} = 1.63, \tilde{I}_{Cl} = 1.89$(nA), and a transient production of a burst of high-frequency, $v = 6$ Hz, action potentials of amplitude 53 mV. The fast T-type Ca^{2+} current provides the main influx of intracellular calcium during which $\max[Ca_i^{2+}] = 0.62$ μM is recorded. The fasciculus generates the active force, $T^a = 20.7$ mN/cm. The reversal of extracellular calcium to its physiological level brings the myofiber to its original electromechanical activity.

Slow wave oscillations cease in a calcium-free environment. The fasciculus becomes hyperpolarized at the constant level, $V = -50$ mV. The concentration of intracellular calcium decreases to 0.1 μM and is insufficient to sustain mechanical contractions. The myofiber remains relaxed.

3.2.3   Changes in K_0^+

A twofold increase in the concentration of extracellular potassium depolarizes the membrane, $V^\tau = -30$ mV, and abolishes slow waves (Fig. 3.3). The ion

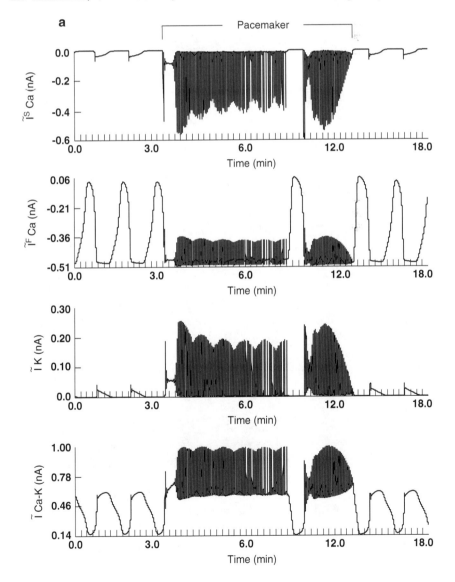

Fig. 3.1 (continued)

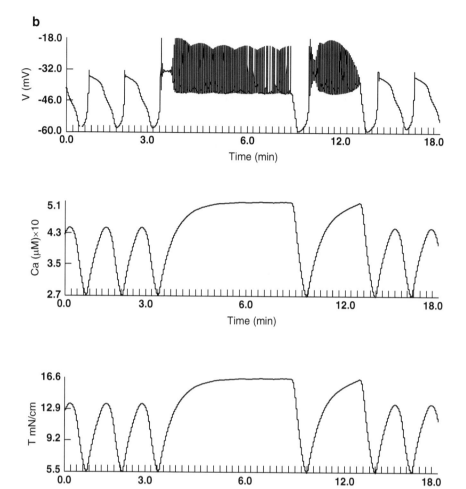

Fig. 3.1 Normal electromechanical activity of the myometrial fasciculus. Traces from *top* to *bottom* indicate: ion currents, depolarization wave dynamics, intracellular calcium changes, and total force

currents display a constant dynamics: $\tilde{I}_{Ca}^s = 0.04$; $\tilde{I}_{Ca}^f = 0.48$; $\tilde{I}_{K1} = 0.42$; $\tilde{I}_{Ca-K} = 0.51\,(nA)$. L- and T-$Ca^{2+}$ currents contribute equally to the rise in the intracellular calcium level, $[Ca_i^{2+}] = 0.5\,\mu M$, and the contraction of the myofiber, $T^a = 15\,mN/cm$.

The following fourfold increase in $[K_0^+]$ further depolarizes the membrane, $V^r = -20\,mV$. The intensity of the outward potassium current increases to $0.52\,nA$ with a concomitant attenuation of the respective currents: $\tilde{I}_{Ca}^s = 0.032$; $\tilde{I}_{Ca}^f = 0.39$ and $\tilde{I}_{Ca-K} = 0.14\,(nA)$. There is an exponential decline in $[Ca_i^{2+}]$ to $0.44\,\mu M$ and in the intensity of force, $T^a = 13.6\,mN/cm$.

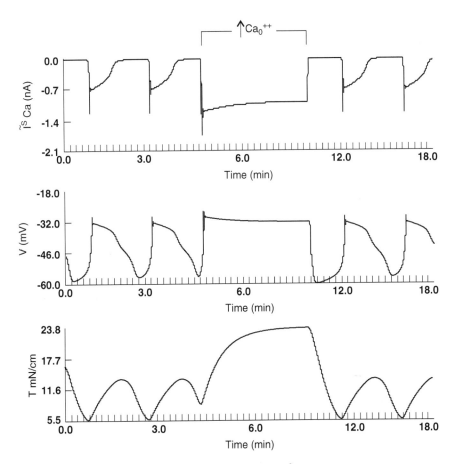

Fig. 3.2 Response of the fasciculus to changes in external Ca^{2+} concentration

A superimposed electrical excitation leads to a burst of high-amplitude, $V = 30$ mV, and frequency, $\nu = 7.3$ Hz, action potentials. The intracellular calcium content rises to 0.52 µM, and the fasciculus produces the active force of 16.8 mN/cm.

A simultaneous elevation in the concentration of extracellular potassium and calcium ions stabilizes the membrane potential at -18 mV. The intracellular calcium, $\left[Ca_i^{2+}\right] = 0.65\,\mu M$, triggers a strong contraction of the myofiber, max $T^a = 21.8$ mN/cm (Fig. 3.4).

A gradual reduction of $\left[K_0^+\right]$ hyperpolarizes the fasciculus, $V^\tau = -70$ and -88 (mV) (Fig. 3.5). Slow wave amplitude and frequency increase to 40 mV, $\nu = 0.032$ Hz and 58 mV, $\nu = 0.037$ Hz. Concurrent multiple discharges of a pacemaker evoke the production of spikes of an average amplitude of 60 mV at a frequency of ~6 Hz. There is a weakening of the calcium influx, max $\left[Ca_i^{2+}\right] = 0.25$ and 0.18 µM, and tension, $T^a = 4.5$ and 2.2 (mN/cm). Interestingly, the

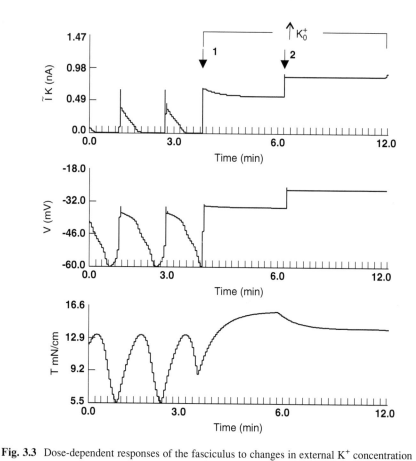

Fig. 3.3 Dose-dependent responses of the fasciculus to changes in external K^+ concentration

electrical stimulation further reduces the strength of contraction, $T^a = 1.1 \, mN/cm$. The duration of contractions also decreases.

3.2.4 Changes in Cl_0^-

A 1.5 times normal decrease in the chloride extracellular concentration lowers the resting membrane potential of the myometrium to -68 mV. Slow waves are generated at a frequency of 0.027 Hz, and the amplitude of 35 mV. There is a decrease in intracellular calcium, 0.33 μM, and an associated fall in the strength of tension, $T^a = 10 \, mN/cm$. A twofold decrease in $\left[Cl_0^- \right]$ hyperpolarizes the fasciculus to -75 mV and ceases its oscillatory activity. The concentration of intracellular calcium drops to 0.03 μM and as such cannot induce spontaneous mechanical contractility (Fig. 3.6).

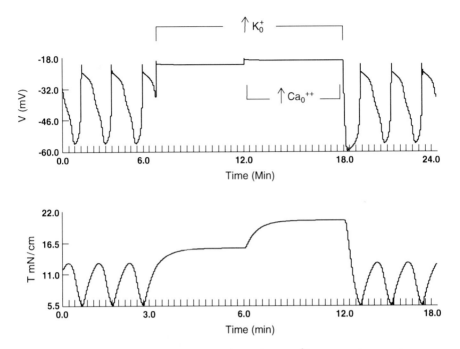

Fig. 3.4 Effect of conjoint changes in extracellular K^+ and Ca^{2+} concentrations on electrome-chanical responses of the myometrial "fiber"

The subsequent elevation of the extracellular chloride concentration abolishes slow waves and depolarizes the membrane at -35 mV. It is associated with a rise in cytosolic calcium to $0.48\,\mu M$ and the development of tonic contraction, $T^a = 16.5\,\text{mN/cm}$.

3.3   Effects of Ion Channel Agonists/Antagonists

3.3.1   T-Type Ca^{2+} Channels

Cumulative addition of a selective T-type Ca^{2+} channel blocker, mibefradil, produces concentration-dependent reduction of the fast inward calcium current, the amplitude and frequency of slow waves, calcium transients, and the intensity of contractions of the myometrium (Fig. 3.7). Thus, immediately after application of the drug, there is a decrease in the amplitude of the fast T-type Ca^{2+} to 0.2 nA. The amplitude and frequency of slow waves diminish to 10 mV and 0.015 Hz, respectively. The concentration of intracellular calcium drops to $0.3\,\mu M$ and the strength of the active force to $12.1\,\text{mN/cm}$. At "high" concentrations, mibefradil abolishes oscillatory electrical and inhibits phasic spontaneous contractions. The membrane

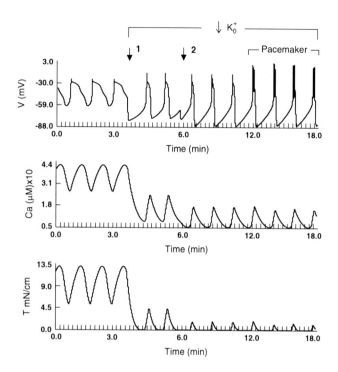

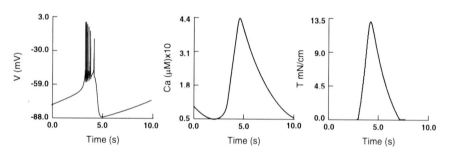

Fig. 3.5 Dose-dependent responses of the fasciculus to gradual changes in external K$^+$ concentration and high-frequency electrical stimulation

becomes depolarized, $V^r = -44$ mV, and the myofiber produces a spastic, tonic-type contraction, $T^a = 11.6$ mN/cm.

Conjoint application of mibefradil and electrical stimulation of the myometrium induce the high-frequency and low-amplitude transient T-type Ca^{2+} current of strength 0.13 nA. There is an increase in $\left[\mathrm{Ca}_i^{2+}\right] = 0.5$ μM and the intensity of contraction, max $T^a = 16.7$ mN/cm. After washout of the drug, the fasciculus does regain its normal physiological activity.

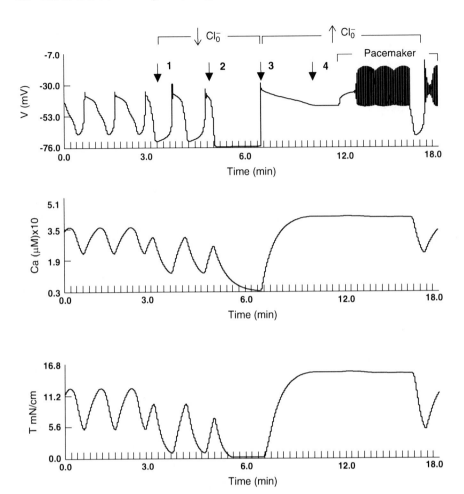

Fig. 3.6 Effects of altered external Cl⁻ ion concentration and pacemaker discharges on electromechanical activity of the fasciculus

3.3.2   *L-Type* Ca^{2+} *Channels*

Nimodipine, a selective L-type Ca^{2+} channel blocker, at "low" doses slightly increases the frequency and decreases the amplitude of slow waves, $V = 19.3$ mV (Fig. 3.8). Reduction in the intensity of the slow inward calcium current to 0.02 nA results in a decrease in free cytosolic calcium, $[Ca_i^{2+}] = 0.31 \mu$M. The strength of contractions also diminishes, $T^a = 7.2$ mN/cm. An increase in the concentration of the drug nearly abolishes the influx of extracellular calcium, $\tilde{I}_{Ca}^s \simeq 0$ nA, and attenuates further the amplitude of slow waves, $V = 12$ mV. However, the myofiber continues to generate phasic contractions of 4.4 mN/cm.

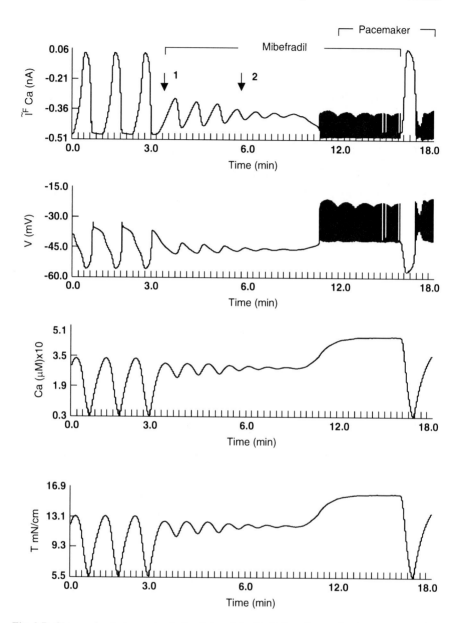

Fig. 3.7 Changes in electromechanical activity of the fasciculus after application of mibefradil – selective T-type Ca^{2+} channel antagonist – and conjoint electrical stimulation

Conjoint electrical discharges of the pacemaker cell in the presence of nimodipine induces high-amplitude action potentials, $V = 38$ mV, on the crests of slow waves. The myometrium contracts with $T^a = 10$ mN/cm. The "high" dose of the drug inhibits spiking but not mechanical activity of the fasciculus. It continues generating the active force of an amplitude of 4.4 mN/cm.

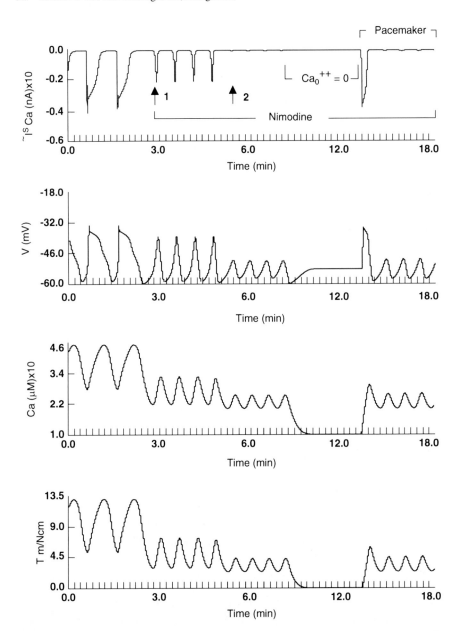

Fig. 3.8 Electromechanical response of the fasciculus to nimodipine, a selective L-type Ca^{2+} channel antagonist, $[Ca_0^{2+}] = 0$ and pacemaker discharges

Nimodipine in a Ca_0^{2+}-free medium completely inhibits electrical activity in the myofiber. It remains relaxed and depolarized, $V = -56$ mV.

Application of nifedipine, a nonselective T- and L-type Ca^{2+} channel antagonist, or a combination of nifedipine and mibefradil, inhibits spontaneous electrical and

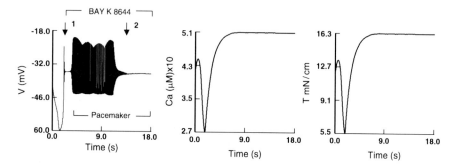

Fig. 3.9   Effect of Bay K 8644 – a Ca^{2+} channel agonist – on biomechanics of the fasciculus

mechanical contractile activity. The intracellular calcium level decreases to 0.03 μM and no contractions are produced.

The response of the myofiber to treatment with Bay K 8644, a weakly selective L-type Ca^{2+} channel agonist, is dose dependent. At "low" concentration, the drug has a depolarizing effect on the myometrium, $V^r = -33$ mV. However, it does not attenuate its excitability. Multiple discharges of pacemaker evoke high-frequency, $v \simeq 6$ Hz, spikes of average amplitude, $V = 35$ mV. The concentration of intracellular calcium rises exponentially to 0.5 μM and is associated with active force production, max $T^a = 16.5$ mN/cm (Fig. 3.9). BAY K 8644 at "high" dose completely inhibits slow waves. The myofiber becomes depolarized and undergoes tonic contraction.

3.3.3 BK_{Ca} Channels

Treatment of the electrically stimulated myometrium with iberiotoxin, a selective maxi Ca^{2+}-activated K^+ channel blocker, exerts a strong excitatory effect on its myoelectrical activity. There is a slight increase in the amplitude of action potentials, $V = 48/50$ mV and the frequency, $v = 3.2$ Hz. The level of free intracellular calcium rises to a maximum of 0.51 μM and the myofiber generates a tonic contraction, max $T^a = 16.2$ mN/cm (Fig. 3.10).

Addition of a "low" dose of nifedipine disrupts the continuous pattern of activity. The myofiber generates regular bursts of spikes of duration of 2 min and of normal amplitude. The contractility pattern changes from a tonic to phasic type. An increase in the concentration of the drug hyperpolarizes the cell membrane, $V^r = -52$ mV, and abolishes its oscillatory activity. Free intracellular calcium of 0.15 μM triggers the active force development of intensity 1 mN/cm. After washout of iberiotoxin and nifedipin, the fasciculus regains its physiological myoelectrical activity.

BMS 191011, a potent selective BK_{Ca} channel agonist, at "low" concentration, causes a slight increase in the frequency of slow waves without changes in their

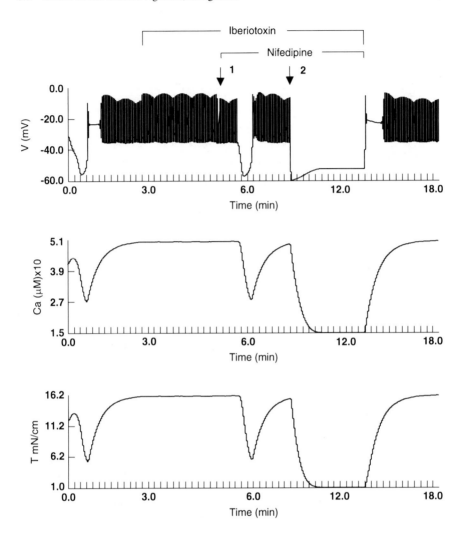

Fig. 3.10 Myoelectrical activity of the myometrial fiber in the presence of iberiotoxin and nifedipine

amplitude. It has a distinct effect on the dynamics of intracellular calcium and contractility. The maximum $\left[Ca_i^{2+}\right]$ declines to 0.35 μM, and the active force to 9.6 mN/cm (Fig. 3.11). At "high" concentration, the compound hyperpolarizes the membrane, $V^r = -66$ mV, and causes a decrease in the amplitude of V to 24 mV. The myofiber continues to produce phasic contractions of strength, $T^a = 6.4$ mN/cm.

BMS 191011 does not inhibit excitability of the myometrium. It generates bursts of action potentials of 50 mV in response to discharges of the intrinsic pacemaker. However, the duration of action potentials decreases with increases in the concentration of the added compound.

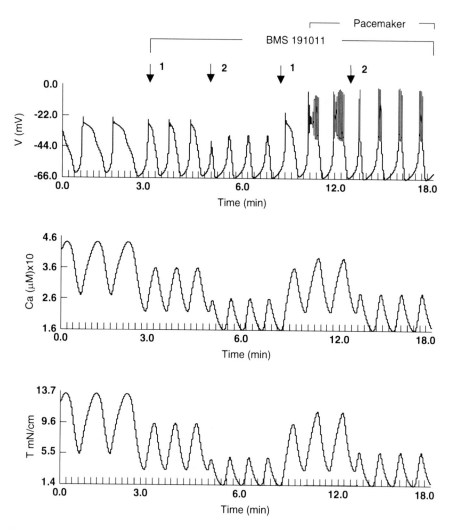

Fig. 3.11 Dose-dependent effects of BMS 191011 – a potent selective BK_{Ca} channel agonist – and high-frequency electrical stimulation on electromechanical activity of the fasciculus

3.3.4 K^+ Channels

Addition of tetraethylammonium chloride (TEA), a nonselective voltage-gated K_{v1}^+ channel antagonist, prolongs the plateau membrane potential duration and increases its amplitude, $V = 28/30\,\mathrm{mV}$. There is an increase in the influx of calcium ions inside the cell, $\left[Ca_i^{2+}\right] = 0.35\ \mu M$, and a rise in the tension of contractions, $T^a = 16\,\mathrm{mN/cm}$. These effects are dose dependent (Fig. 3.12).

Electrical stimulation of the fasciculus in the presence of the drug at "low" concentration inhibits slow waves and depolarizes the membrane, $V = -26$ mV.

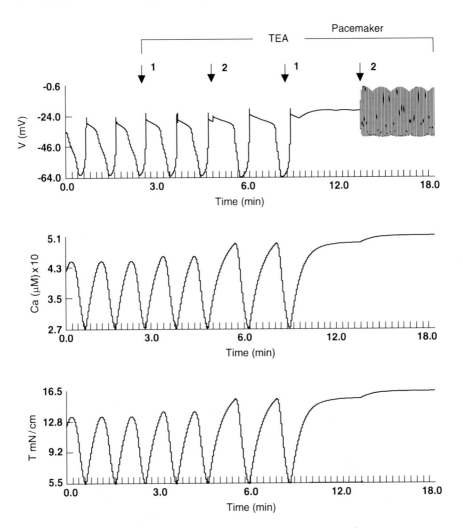

Fig. 3.12 Effects of tetraethylammonium chloride (TEA) – a selective K$^+$ channel agonist – applied in gradually increased concentrations on biomechanics of the fasciculus

These events coincide with an increase in cytosolic $\left[Ca_i^{2+}\right] = 0.48\,\mu M$ and the production of tonic contraction. At "high" concentrations, TEA evokes long-lasting bursts of regular action potentials, $V = 25/27$ mV. The myometrium generates the active force of a maximum of 16.5 mN/cm.

3.3.5 Cl$^-$ *Channels*

Niflumic acid, a nonselective Cl$^-$-channel antagonist and a potent stimulator of the BK$_{Ca}$ channel, at "low" concentrations reduces the strength of the leak chloride

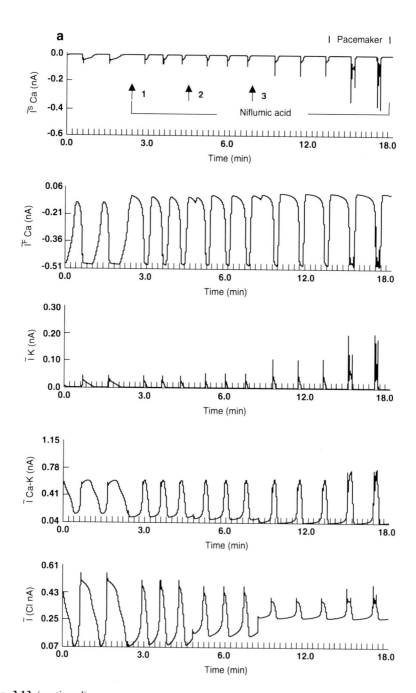

Fig. 3.13 (continued)

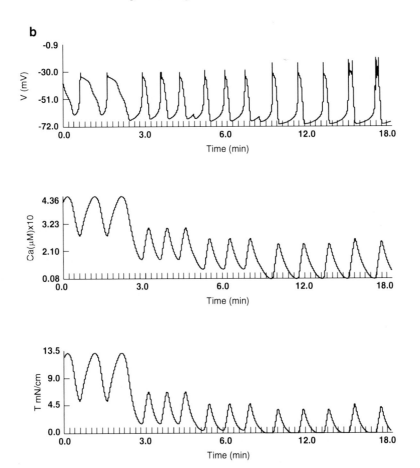

Fig. 3.13 Dose-dependent responses of the fasciculus to niflumic acid at different concentrations and pacemaker discharges

current to 0.4 nA without significant changes in intensities of other ion currents (Fig. 3.13). At "high" dose, the drug causes a rise in amplitudes of slow inward calcium and potassium currents to 0.2 nA and 0.11 nA, respectively, while the amplitude of the chloride current diminishes to 0.15 nA. The membrane becomes hyperpolarized, $V^r = -70$ mV. However, the myometrium continues to generate slow waves of amplitude ~40 mV at a frequency of 0.44 Hz. Electrical stimuli induce action potentials, $V = -10$ mV, on the crests of slow waves.

A gradual increase in the concentration of niflumic acid causes a decrease in the intracellular concentration of free calcium to 3 μM at "low" dose, and to 2.2 μM at "high" dose. The drug at "high" dose causes further decline in cytosolic calcium, $[Ca_i^{2+}] = 0.08 \, \mu M$. This significantly weakens contractility of the myofiber that generates the active force of 4.3 mN/cm. High-intensity electrical stimulation does not have any positive effect on the dynamics of contractions.

Exercises

1. Use ABS Technologies® software to simulate the effect of high-frequency stimulation on electromechanical activity of the myofiber.
2. Use ABS Technologies® software to study the effects of myometrial stiffening/softening on stress–strain development in the myofiber.
3. Use ABS Technologies® software to simulate the effects of long-standing hyperglycemia on electromechanical activity of the myofiber.
4. Results of pharmacological experiments obtained on isolated ion channels can lead to erroneous conclusions regarding how the system of multiple channels works. Use ABS Technologies® software to analyze the effects of a selective ion channel agonist/antagonist on the dynamics of other membrane ion channels present in the cell.
5. Use ABS Technologies® software to study the effect of prestretching of the myofiber on its electromechanical activity.
6. Benzodiazpines are drugs that are commonly administered to treat anxiety before labor. Two types, i.e., A and B, of GABA receptors have been identified in human myometrium. Use ABS Technologies® software to study the effect of benzodiazepine (Midazolam) on contractility of the myofiber (Hint: GABA A-B receptors are coupled to ligand-gated L-type Ca^{2+} and Cl^- channels).
7. Various conceptual mathematical frameworks are used to combine observations across multiple space and time scales. Bursztyn et al. (2007) proposed a morphostructural model of a human myometrial cell. Discuss advantages and disadvantages of Bursztyn's model with regard to the phenomenological model developed in the book.

Chapter 4
General Theory of Thin Shells

Everything should be made as simple as possible, but not one bit simpler.

A. Einstein

4.1 Basic Assumptions

A solid body bounded by two closely spaced curved surfaces is called a shell. Consider a surface S within the shell. Any point M on S can be associated with general curvilinear coordinates a^1, a^2 and the unit normal vector $\bar{m}$. Let the perpendicular distance along $\bar{m}$ be given by z: $-0.5h$ $(a^1, a^2 \leq z \leq 0.5h)$ (a^1, a^2), where h is the thickness of the shell. The surface S for which $z = 0$ is called the middle surface. The two surfaces defined by $z = \pm 0.5h (a^1, a^2)$ are called the faces of the shell. Throughout this book we assume that the faces of the shell are smooth and contain no singularities, i.e., there are no structural defects or inclusions in the wall of the gravid uterus, and the thickness h is constant (Fig. 4.1).

The shell is classified as thin if the $\max(h/R_i) \leq 1/20$, where R_i are the radii of the curvature of S. However, the estimate is very rough and in various practical applications other geometric and mechanical characteristics should also be considered.

The following paragraphs assemble all the required preliminary results from the theory of surfaces which are required to understand and develop the concepts of shells (Galimov 1975). The reader familiar with the subject of the matter can proceed directly to Chap. 5.

R.N. Miftahof and H.G. Nam, *Biomechanics of the Gravid Human Uterus*, DOI 10.1007/978-3-642-21473-8_4, © Springer-Verlag Berlin Heidelberg 2011

Fig. 4.1 A thin shell. Used
with permission from
Cambridge University Press

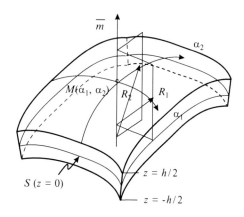

4.2 Geometry of the Surface

Let the middle surface S of the bioshell be associated with a set of independent
parameters a^1 and a^2 (Fig. 4.2). Putting $a^1 = $ const and varying the parameter a^2 we
obtain a curve that lies entirely on S. Successively giving a^1 a series of constant
values we obtain a family of curves along which only a parameter a^2 varies. These
curves are called the a^2 - coordinate lines. Similarly, setting $a^2 = $ const we obtain
the a^1 - coordinate lines of S. We assume that only one curve of the family passes
through a point of the given surface. Thus, any point M on S can be treated as a
cross-intersection of the a^1- and a^2 - curvilinear coordinate lines.

The position of a point M with respect to the origin O of the reference system is
defined by the position vector $\bar{r}$. Differentiating $\bar{r}$ with respect to a^i ($i = 1, 2$)
vectors tangent to the a^1- and a^2 - coordinate lines can be found

$$\bar{r}_1 = \frac{\partial \bar{r}}{\partial \alpha^1}, \quad \bar{r}_2 = \frac{\partial \bar{r}}{\partial \alpha^2}. \tag{4.1}$$

The vector $\bar{m}$ normal to $\bar{r}_1$ and $\bar{r}_2$ is found from

$$\bar{m} = [\bar{r}_1, \bar{r}_2] \quad \text{and} \quad \bar{m}\bar{r}_1 = 0, \quad \bar{m}\bar{r}_2 = 0, \tag{4.2}$$

where $[\bar{r}_1, \bar{r}_2]$ is the vector product. The vectors $\bar{r}_1, \bar{r}_2$, and $\bar{m}$ are linearly independent and comprise a covariant base $\{\bar{r}_1, \bar{r}_2, \bar{m}\}$ on S. We introduce the unit vectors $\bar{e}_i$
in the direction of $\bar{r}_i$ as

$$\bar{e}_1 = \frac{\bar{r}_1}{|\bar{r}_1|} = \frac{\bar{r}_1}{A_1}, \quad \bar{e}_2 = \frac{\bar{r}_2}{|\bar{r}_2|} = \frac{\bar{r}_2}{A_2}, \tag{4.3}$$

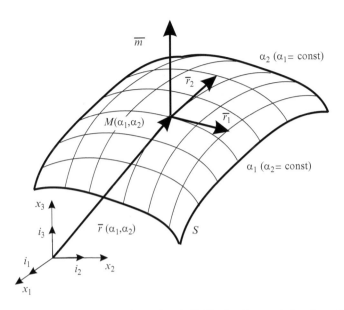

Fig. 4.2 Parameterization of the surface of a shell. Used with permission from Cambridge University Press

where

$$|\bar{r}_1| = \bar{r}_1\bar{r}_1 = A_1 = \sqrt{a_{11}}, \quad |\bar{r}_2| = \bar{r}_2\bar{r}_2 = A_2 = \sqrt{a_{22}},$$
$$\bar{r}_1\bar{r}_2 = A_1A_2 \cos \chi = \sqrt{a_{12}}. \tag{4.4}$$

Here χ is the angle between coordinate lines, A_i are the Lamé parameters.

Covariant components of the first $\mathbf{A}(a_{ik})$ and second $\mathbf{B}(b_{ik})$ metric tensors on S are defined by

$$a_{ik} = \bar{r}_i\bar{r}_k, \quad b_{ik} = -\bar{r}_i\bar{m}_k = -\bar{r}_k\bar{m}_i = \bar{r}_{ik}\bar{m}, \quad (i,k = 1,2) \tag{4.5}$$

Mixed components of the second metric tensor are $b_k^i = a^{is}b_{ks}$. Similarly, covariant (c_{ik}) and contravariant (c^{ik}) components of the discriminant tensor $\mathbf{C}$ are given by

$$\bar{m}[\bar{r}_i, \bar{r}_k] = c_{ik}, \quad \bar{m}[\bar{r}^i, \bar{r}^k] = c^{ik}. \tag{4.6}$$

Hence,

$$c_{ik}\bar{m} = [\bar{r}_i, \bar{r}_k], \quad c^{ik}\bar{m} = [\bar{r}^i, \bar{r}^k],$$
$$c_{ik}\bar{r}^k = [\bar{m}, \bar{r}_i], \quad c^{ik}\bar{r}_i = [\bar{m}, \bar{r}^i], \tag{4.7}$$

it follows that

$$c^{ii} = 0, \quad c_{ii} = 0, \quad c^{12} = -c^{12} = \frac{1}{\sqrt{a}}, \quad c_{12} = -c_{12} = \sqrt{a},$$
$$c_{ik}c^{km} = \delta^m{}_i, \quad c_{ik}c^{ki} = \delta^i_i = 2. \tag{4.8}$$

Here forward the summation convention is applied, i.e., when an index variable appears twice in a single term, once in an upper (superscript) and once in a lower (subscript) position, it implies that we are summing over all of its possible values.

Vectors $\vec{r}^i$ are reciprocal to $\vec{r}_i$ $(\vec{r}^i = a^{ik}\vec{r}_k)$ such that

$$\vec{r}^k\vec{r}_i = \delta^k_i, \quad \vec{r}^k\vec{m} = 0, \quad [\vec{r}^k, \vec{m}] = c_{ik}\vec{r}^k, \quad \vec{r}^i\vec{r}^k = a^{ik}, \tag{4.9}$$

here δ^k_i is the Kronecker delta

$$\delta^k_i = \begin{cases} 1, & \text{if } i = k \\ 0, & \text{if } i \neq k. \end{cases}$$

Covariant derivatives of a vector or a tensor are found to be

$$\nabla_k a_i = \partial_k \alpha_i - \Gamma^j_{ik}\alpha_j, \quad \nabla_k a^i = \partial_k \alpha^i + \Gamma^i_{jk}\alpha^j,$$
$$\nabla_j T^{ik} = \partial_j T^{ik} + T^{sk}\Gamma^i_{js} + T^{is}\Gamma^k_{js}, \quad \nabla_j T_{ik} = \partial_j T_{ik} - T_{ks}\Gamma^s_{ij} - T_{is}\Gamma^s_{jk}, \tag{4.10}$$

where Γ^j_{ik} ($\Gamma^j_{ik} = \vec{r}^j\vec{r}_{ik}$) are the Christoffel symbols of the second kind, ∇_k is the sign of covariant differentiation with respect to a_{ik}, $\partial_k \alpha_i = \partial\alpha^k$.

The following Gauss-Weingarten equations hold

$$\vec{r}_{ik} = \Gamma^j_{ik}\vec{r}_j + \vec{m}b_{ik}, \qquad \vec{m}_i = \partial_i\vec{m} = -b^k_i\vec{r}_k,$$

or

$$\nabla_i\vec{r}_k = \vec{m}b_{ik}, \quad \nabla_i\vec{r}^k = \vec{m}b^k_i, \quad \nabla_i\vec{m} = -b^k_i\vec{r}_k. \tag{4.11}$$

Note that the above formulae are also valid for the deformed surface $\overset{*}{S}$ (Henceforth, all quantities that refer to the deformed configuration we shall designate by an asterisk (*) unless otherwise specified).

Let u_i and u^i be co- and contravariant components of the vector of displacement $\vec{v}(\alpha^1, \alpha^2)$ referred to the deformed coordinate system on $\overset{*}{S}$. Thus

$$\overset{*}{\vec{r}} = \vec{r} + \vec{v}(\alpha^1, \alpha^2), \quad \vec{v} = u_i\vec{r}^i + \vec{m}w = u^i\vec{r}_i + \vec{m}w, \tag{4.12}$$

where $\overset{*}{\bar{r}}$ is the position vector of a point M on $\overset{*}{S}$, u_i, u^i are the tangent, and w is the normal displacements (deflection), respectively. Covariant differentiation of (4.12) with respect to α^i for coordinate vectors $\overset{*}{\bar{r}}_i$, we find

$$
\begin{aligned}
\overset{*}{\bar{r}}_i &= \bar{r}_k(\delta_i^k + e_i^k) + \bar{m}\omega_i, \\
e_i^k &= \nabla_i u^k - w b_i^k, \qquad \omega_i = \nabla_i w + b_{ik}u^k.
\end{aligned}
\tag{4.13}
$$

The covariant components ε_{ik} of the tensor of deformations $\mathbf{E}$ on S are defined by

$$
2\varepsilon_{ik} = \overset{*}{a}_{ik} - a_{ik} = \overset{*}{\bar{r}}_i\,\overset{*}{\bar{r}}_k - \bar{r}_i\bar{r}_k = e_{ik} + e_{ki} + a^{js}e_{ij}e_{ks} + \omega_i\omega_k,
\tag{4.14}
$$
$$
e_{ik} = \nabla_i u_k - w b_{ik}, \quad (i,k = 1,2).
$$

Here use is made of (4.13) for $\overset{*}{\bar{r}}_i$. The components $\varepsilon_{11}, \varepsilon_{22}$ describe tangent deformations along the coordinate lines α^1, α^2 and ε_{12} is the shear deformation that characterizes the change in the angle between them.

The vector $\overset{*}{\bar{m}}$ $(\overset{*}{\bar{m}} \perp \overset{*}{S})$ is found from

$$
\overset{*}{\bar{m}} = \frac{(\bar{r}^i\mathbb{S}_i + \bar{m}\mathbb{S}_3)}{\sqrt{\mathfrak{A}}},
\tag{4.15}
$$

where the following notations are introduced

$$
\begin{aligned}
\mathbb{S}_i &= e_i^k\omega_k - (1 + e_1^1 + e_2^2)\omega_k, \quad \mathbb{S}_3 = (1 + e_1^1)(1 + e_2^2) - e_2^1 e_1^2, \\
\mathfrak{A} &= 1 + 2\varepsilon_i^i + 4(\varepsilon_1^1\varepsilon_2^2 - \varepsilon_1^2\varepsilon_2^1)
\end{aligned}
\tag{4.16}
$$

The quantities $e_{ik}, v_i, \mathbb{S}_i, \mathbb{S}_3$ describe the rotations of tangent and normal vectors during deformation.

Covariant components of the tensor of bending deformations on S defined by

$$
\mathfrak{æ}_{ik} = \overset{*}{b}_{ik} - b_{ik}, \quad (b_{ik} = -\bar{m}_i\bar{r}_k, \overset{*}{b}_{ik} = -\overset{*}{\bar{m}}_i\,\overset{*}{\bar{r}}_k),
\tag{4.17}
$$

where $\mathfrak{æ}_{ii}$ describe changes in curvatures, and $\mathfrak{æ}_{12}$ is the twist of the surface. Substituting $\overset{*}{\bar{r}}_k$ and $\overset{*}{\bar{m}}_k = \partial_k\overset{*}{\bar{m}}$ from (4.13) and (4.15) into the above, we find

$$
\overset{*}{b}_{ik} = -\frac{1}{\sqrt{\mathfrak{A}}}\left(\omega_k\nabla_i\mathbb{S}_3 + (\delta_k^j + e_k^j)\nabla_i\mathbb{S}_j - b_{ij}\left(\mathbb{S}_3(\delta_k^j + e_k^j) - \omega_k\mathbb{S}^j\right)\right).
\tag{4.18}
$$

Making use of the right-hand sides of equalities

$$
\omega_k\nabla_i\mathbb{S}_3 + (\delta_k^j + e_k^j)\nabla_i\mathbb{S}_j = -\mathbb{S}_3\nabla_i\omega_k - \mathbb{S}_j\nabla_i e_k^j,
$$

$$(\delta^n_j + e^n_j)\nabla \mathbb{S}_3 - \omega_j \mathbb{S}^n = \overset{*}{a}_{js}\big((2 + e^1_1 + e^2_2)a^{sn} - (\delta^s_k + e^s_k)a^{kn}\big),$$

Equation (4.18) takes the final form

$$\overset{*}{b}_{ik} = \frac{1}{\sqrt{\mathfrak{A}}}\Big[\mathbb{S}_3\nabla_i\omega_k + \mathbb{S}_j\nabla_i e^j_k + b_{ij}\overset{*}{a}_{js}\big((2 + e^1_1 + e^2_2)a^{js} - (\delta^s_t + e^s_t)a^{jt}\big)\Big]. \quad (4.19)$$

4.3 Tensor of Affine Deformation

The Christoffel symbols of the first kind of the undeformed and deformed middle surfaces are defined by

$$\overset{(*)}{\Gamma}_{j,ik} = \frac{1}{2}\left(\partial_k \overset{(*)}{a}_{ij} + \partial_i \overset{(*)}{a}_{kj} - \partial_j \overset{(*)}{a}_{ik}\right) \quad (i, j, k = 1, 2). \quad (4.20)$$

Substituting $\overset{*}{a}_{ik} = a_{ik} + 2\varepsilon_{ik}$ from (4.14) into (4.20), we get

$$\overset{*}{\Gamma}_{j,ik} = \Gamma_{j,ik} + \partial_k\varepsilon_{ij} + \partial_i\varepsilon_{jk} - \partial_j\varepsilon_{ik}. \quad (4.21)$$

The right-hand side of (4.21) can be written in the form

$$\partial_k\varepsilon_{ij} + \partial_i\varepsilon_{jk} - \partial_i\varepsilon_{ik} = P_{j,ik} + 2\,\Gamma^n_{ik}\varepsilon_{nj},$$

where

$$P_{j,ik} = \nabla_i\varepsilon_{jk} + \nabla_k\varepsilon_{ij} - \nabla_j\varepsilon_{ik}, \quad (4.22)$$

and

$$\nabla_j\varepsilon_{ik} = \partial_j\varepsilon_{ik} - \Gamma^n_{ij}\varepsilon_{nk} - \Gamma^n_{kj}\varepsilon_{in}, \quad \Gamma^n_{ik} = a^{nj}\,\Gamma_{j,ik}.$$

Expression (4.21) can be written in the form

$$\overset{*}{\Gamma}_{j,ik} = \Gamma_{j,ik} + P_{j,ik} + 2\,\Gamma^n_{ik}\varepsilon_{nj},$$

which upon contracting of both sides with the tensor $\overset{*}{a}^{js}$ yields

$$\overset{*}{\Gamma}^j_{ik} = \Gamma^j_{ik} + A^j_{ik} \quad (4.23)$$

where $A_{ik}^{j} = a^{jn} \overset{*}{P}_{n,ik}$. Note, that although the Christoffel symbols are not tensors their difference, A_{ik}^{j}, is a tensor of the third valency such that $A_{ik}^{j} = A_{ki}^{j}$. It tensor is called the tensor of affine deformation (Norden 1950).

The tensor A_{ik}^{j} allows to establish a relationship between covariant derivatives of a vector A^{k} with respect to $\overset{*}{a}_{ik}$ and a_{ik}

$$\overset{*}{\nabla}_i A^k = \nabla_i A^k + A_{ij}^k A^j, \quad \overset{*}{\nabla}_i A_k = \nabla_i A_k - A_{ik}^j A_j. \tag{4.24}$$

Similar relationships can be set for tensors of any valency.

Consider coordinate vectors $\bar{r}_k$ as tensors of first valency. Their covariant derivatives are given by

$$\overset{*}{\nabla}_i \bar{r}_k = \bar{r}_{ik} - \overset{*}{\Gamma}_{ik}^s \bar{r}_s,$$

here $\overset{*}{\nabla}_i$ – the sign of covariant differentiation in the metric $\overset{*}{a}_{ik}$. With the help of (4.11) we get

$$\overset{*}{\bar{r}}_{ik} = \overset{*}{\Gamma}_{ik}^s \overset{*}{\bar{r}}_s + \overset{*}{b}_{ik} \overset{*}{\bar{m}}, \quad \overset{*}{\bar{m}}_i = -\overset{*}{b}_i^k \overset{*}{\bar{r}}_k. \tag{4.25}$$

The following is true

$$\overset{*}{\nabla}_i \overset{*}{\bar{r}}_k = \overset{*}{b}_{ik} \overset{*}{\bar{m}}, \quad \overset{*}{\nabla}_i \overset{*}{\bar{m}} = -\overset{*}{b}_i^k \overset{*}{\bar{r}}_k.$$

Substituting (4.23) into the above, we obtain

$$\overset{*}{\nabla}_i \overset{*}{\bar{r}}_k = \overset{*}{b}_{ik} \overset{*}{\bar{m}} = \bar{r}_{ik} - \Gamma_{ik}^s \overset{*}{\bar{r}}_s + A_{ik}^s \overset{*}{\bar{r}}_s,$$

from where it follows

$$\nabla_i \overset{*}{\bar{r}}_k = A_{ik}^s \overset{*}{\bar{r}}_s + \overset{*}{b}_{ik} \overset{*}{\bar{m}}, \quad \nabla_i \overset{*}{\bar{m}} = \overset{*}{\nabla}_i \overset{*}{\bar{m}} = -\overset{*}{b}_i^k \overset{*}{\bar{r}}_k. \tag{4.26}$$

Here use is made of (4.24).

For vectors of the reciprocal basis, $\overset{*}{\bar{r}}^k$, we have

$$\overset{*}{\nabla}_i \overset{*}{\bar{r}}^k = \bar{r}_i^k - \overset{*}{\Gamma}_{is}^k \overset{*}{\bar{r}}^s, \quad \bar{r}_i^k = \overset{*}{\Gamma}_{is}^k \overset{*}{\bar{r}}^s + \overset{*}{b}_i^k \overset{*}{\bar{m}}.$$

Substituting $\bar{r}_i^k$ into the first expression, we find: $\overset{*}{\nabla}_i \overset{*}{\bar{r}}^k = \overset{*}{b}_i^k \overset{*}{\bar{m}}$. Making use of (4.23), the final formulars are found to be

$$\nabla_i \overset{*}{\bar{r}}^k = -A_{ij}^k \overset{*}{\bar{r}}^j + \overset{*}{b}_i^k \overset{*}{\bar{m}}, \quad \overset{*}{\bar{m}}_i = -\overset{*}{b}_{ik} \overset{*}{\bar{r}}^k. \tag{4.27}$$

Multiplying (4.26) and (4.27) by $\overset{*}{\bar{r}}_s$, we get

$$A_{ij}^k = \overset{*}{r}{}^k \overset{*}{\nabla}_i \overset{*}{r}_j, \quad A_{ij}^k = - \overset{*}{r}_j \overset{*}{\nabla}_i \overset{*}{r}{}^k. \tag{4.28}$$

Contracting both sides of the fist equality with $\underset{nk}{\overset{*}{a}}$, we have

$$P_{n,ik} = \overset{*}{r}_n \overset{*}{\nabla}_i \overset{*}{r}_k. \tag{4.29}$$

To express the tensor $P_{n,ik}$ in terms of displacements we proceed from differentiating of (4.13) and making use of (4.25) for the undeformed surface S. As a result we obtain

$$\nabla_i \overset{*}{r}_k = (\delta_k^s + e_k^s)\nabla_i \overset{-}{r}_s + \overset{-}{r}_s \nabla_i e_k^s + \overset{-}{m}_i \omega_k + \overset{-}{m}\nabla_i \omega_k$$

$$= \overset{(1)}{W_{ik}^s} \overset{-}{r}_s + \overset{-}{m}_i (\overset{(2)}{W_{ik}} + b_{ik}), \tag{4.30}$$

where

$$\overset{(1)}{W_{ik}^s} = \nabla_i e_k^s - b_k^s \omega_s, \quad \overset{(2)}{W_{ik}} = \nabla_i \omega_k + b_{is} e_k^s.$$

Substituting the resultant equation into (4.29), we have

$$P_{n,ik} = \omega_n (\overset{(2)}{W_{ik}} + b_{ik}) + \overset{(1)}{W_{ik}^s} (a_{ns} + e_{ns}). \tag{4.31}$$

On introducing the right-hand side of (4.30) into the obvious equality $c^{ik}\nabla_i \overset{*}{r}_k = 0$, we obtain

$$c^{ik} \left[\overset{(1)}{W_{ik}^s} \overset{-}{r}_s + \overset{-}{m}_i (\overset{(2)}{W_{ik}} + b_{ik}) \right] = 0.$$

It is clear that the tensors W_{ik}^s, H_{ik} are symmetrical

$$c^{ik} \overset{(1)}{W_{ik}^s} = 0, \quad c^{ik} \overset{(2)}{W_{ik}} = 0, \tag{4.32}$$

or in expanded form

$$c^{ik}\nabla_i e_k^s = c^{ik} b_k^s \omega_i, \quad c^{ik}\nabla_i \omega_k = -c^{ik} b_{is} e_k^s$$
$$c^{ik}\nabla_i e_{ks} = c^{ik} b_{ks}\omega_i, \quad c^{ik}\nabla_i \omega_k = -c^{ik} b_i^s e_{ks}.$$

The last two formulars are obtained by contracting the first two with a_{js}.

The scalar multiplication of the first equation (4.26) by $\overset{*}{m}$, yields $\overset{*}{b}_{ik} = \overset{*}{m} \nabla_i \overset{*}{r}_k$. On substituting $\overset{*}{m}$ from (4.15) and $\nabla_i \overset{*}{r}_k$ from (4.30), for the coefficients $\overset{*}{b}_{ik}$ of the deformed surface $\overset{*}{S}$, we obtain

$$
\overset{*}{b}_{ik} = \frac{1}{\sqrt{\mathfrak{A}}} \left(\mathbb{S}_3 (\overset{(2)}{W}_{ik} + b_k) + \mathbb{S}_j \overset{(1)}{W}_{ik}^{\,j} \right). \tag{4.33}
$$

4.4   Equations of Continuity of Deformations

For the surface to retain continuity during deformation, the parameters ε_{ik} and α_{ik} $(i, k = 1, 2)$ must satisfy the equations of continuity of deformations. To derive them we proceed from covariant differentiation of (4.26) with respect to α^n

$$
\nabla_n \nabla_i \overset{*}{r}^s = \overset{*}{m} \nabla_n \overset{*}{b}_{ik} + \overset{*}{m}_n \overset{*}{b}_{ik} + P_{s,ik} \nabla_n \overset{*}{r}^s + \overset{*}{r}^s \nabla_n P_{s,ik}.
$$

Substituting expressions for $\nabla_n \overset{*}{r}^s$ and $\overset{*}{m}$ given by (4.13), we find

$$
\nabla_n \nabla_i \overset{*}{r}^s = \overset{*}{m} \nabla_n \overset{*}{b}_{ik} - \overset{*}{b}_{ik} \overset{*}{b}_{ns} \overset{*}{r}^s - A^s_{nj} P_{s,ik} \overset{*}{r}^j + \overset{*}{b}^s_n P_{s,ik} \overset{*}{m} + \overset{*}{r}^s \nabla_n P_{s,ik}.
$$

Multiplying the above by $\overset{*}{r}_j$ and $\overset{*}{m}$, we obtain two scalar equations

$$
\overset{*}{r}_j \nabla_n \nabla_i \overset{*}{r}_k = -\overset{*}{b}_{ik} \overset{*}{b}_{jk} - A^s_{ik} P_{s,jn} + \nabla_n P_{j,ik} \tag{4.34}
$$

$$
\overset{*}{m} \nabla_n \nabla_i \overset{*}{r}_k = \nabla_n \overset{*}{b}_{ik} + \overset{*}{b}^j_n P_{j,ik}. \tag{4.35}
$$

On substituting $\nabla_n P_{j,ik}$ from (4.34) into the equality

$$
\nabla_n \left(P_{j,ik} + P_{i,jk} \right) = 2 \nabla_n \nabla_k \varepsilon_{ij}
$$

and contracting the resultant equation with the tensor $c^{in} c^{jk}$, we get

$$
c^{in} c^{jk} \left(-2 \nabla_n \nabla_k \varepsilon_{ij} + \overset{*}{b}_{ik} \overset{*}{b}_{nj} + A^s_{ik} P_{s,ij} + \overset{*}{r}_j \nabla_n \nabla_i \overset{*}{r}_k \right). \tag{4.36}
$$

In deriving (4.36) use is made of equalities

$$
c^{jk} \overset{*}{r}_i \nabla_n \nabla_j \overset{*}{r}_k = \overset{*}{r}_i \nabla_n c^{jk} \nabla_j \overset{*}{r}_k = 0,
$$

that follow from the requirements of integrability of $c^{jk} \nabla_j \overset{*}{\bar{r}}_k = 0$. The last term in (4.36) can be recast using the Ricci equality given by

$$2c^{in} \nabla_n \nabla_j \overset{*}{\bar{r}}_k = c^{in} R^s_{nik} \cdot \overset{*}{\bar{r}}_s, \qquad (4.37)$$

where R^s_{nik} are the mixed components of the Riemann-Christoffel tensor of the surface defined by

$$R^s_{nik} = b_{nk} b^s_i - b_{ik} b^s_n. \qquad (4.38)$$

Thus, we have

$$\overset{*}{m} c^{in} \nabla_n \nabla_i \overset{*}{\bar{r}}_k = 0, \qquad (4.39)$$

$$
\begin{aligned}
c^{in} c^{jk} \overset{*}{\bar{r}}_j \nabla_n \nabla_i \overset{*}{\bar{r}}_k &= \frac{1}{2} c^{in} c^{jk} R^s_{nik} \overset{*}{a}_{js} \\
&= \frac{1}{2} c^{in} c^{jk} (b_{nk} b_{ij} - b_{ik} b_{nj}) + c^{in} c^{jk} R^s_{nik} \cdot \varepsilon_{js}.
\end{aligned}
\qquad (4.40)
$$

Substituting the right-hand side of (4.35) into (4.39), the Gauss-Codazzi equations for the deformed surface are found to be

$$c^{in} (\nabla_n \overset{*}{b}_{ik} + \overset{*}{b}{}^j_n P_{j,ik}) = 0. \qquad (4.41)$$

Analogously, substituting (4.36) into (4.40), we find a relationship between the coefficients of the first and second fundamental forms of the deformed surface

$$c^{in} c^{jk} \left(-2 \nabla_n \nabla_k \varepsilon_{ij} + \overset{*}{b}_{ik} \overset{*}{b}_{nj} - b_{ik} b_{nj} + A^s_{ik} P_{s,ij} + R^s_{nik} \cdot \varepsilon_{js} \right) = 0. \qquad (4.42)$$

Here use is made of the obvious fact that

$$c^{in} c^{jk} b_{nk} b_{ij} = -c^{in} c^{jk} b_{ik} b_{nj}.$$

Applying the Bianchi formular to the last term of (4.42), we find

$$R^s_{nik} = K c_{ni} c_{ks}, \qquad (4.43)$$

where K is the Gaussian curvature of S defined by

$$K = b^1_1 b^2_2 - b^1_2 b^2_1. \qquad (4.44)$$

Since

$$c^{in}c_{ni} = -\delta_i^i = 2, \quad c^{jk}c_k^s\varepsilon_{jk} = c^{jk}c_{ks}\varepsilon_j^s = -\delta_s^j\varepsilon_j^s = -\varepsilon_j^j,$$

then for $R_{nik}^s \cdot \varepsilon_{js}$ we have

$$c^{in}c^{jk}R_{nik}^s \cdot \varepsilon_{js} = Kc^{in}c^{jk}c_{ni}c_k^s\varepsilon_{js} = -K\delta_i^j c^{jk}c_k^s\varepsilon_{js} = 2K\varepsilon_i^i = 2K(\varepsilon_1^1 + \varepsilon_2^2). \quad (4.45)$$

Introducing (4.45) into (4.42) and remembering (4.17) one of the equations of continuity of finite deformations is found to be

$$c^{in}c^{jk}\left(\nabla_n\nabla_k\varepsilon_{ij} - \frac{1}{2}\mathscr{a}_{ik}\mathscr{a}_{nj} - \frac{1}{2}A_{ik}^s P_{s,in}\right) - \left(2Ha^{ik} - b^{ik}\right)\mathscr{a}_{ik} - Ka^{ik}\varepsilon_{ik} = 0, \quad (4.46)$$

where H is the mean curvature of S and the following formulas hold

$$c^{in}c^{jk}b_{nk} = 2Ha^{ij} - b^{ij}, \quad 2H = a^{ik}b_{ik} = b_i^i.$$

With the help of the Codazzi conditions for the undeformed surface: $c^{in}\nabla_n b_{ik} = 0$, and equalities: $A_{ik}^j = b_n^j P_{j,ik}$, two additional equations of compatibility of deformations can be obtained from (4.41) in the form

$$c^{in}\nabla_n\mathscr{a}_{ik} = c^{in}(b_{jn} - \mathscr{a}_{jn})A_{ik}^j. \quad (4.47)$$

Contracting both sides of (4.47) with the tensor c^{jk}, we get

$$c^{in}c^{jk}\nabla_n\mathscr{a}_{ik} = c^{in}c^{jk}(b_{ns} - \mathscr{a}_{ns})A_{ik}^s. \quad (4.48)$$

4.5   Equations of Equilibrium

Although a thin shell can be treated as a three-dimensional solid the complexity of the problem would be reduced significantly if its dimensionality could be reduced from three to two. Applying the second Kirchhoff-Love hypothesis that "the transverse normal stress is significantly smaller compared to other stresses in the shell and thus may be neglected", and recalling that the deformed state of the shell is completely defined in terms of deformations and curvatures of its middle surface, the shell can be regarded as a two-dimensional solid. Furthermore, restricting our considerations to the membrane theory of shells, i.e., the lateral forces and moments acting upon the element are assumed equal zero, the stress state of a differential element of the membrane is described entirely by in-plane tangent $T_{ii}(T^{ii})$ and shear $T_{ik}(T^{ik})(i \neq k)$ forces per unit length of the element.

To derive the equations of equilibrium of a shell in terms of the deformed configuration, we proceed from the vector equations of equilibrium for a three dimensional solid given by Galimov (1975)

$$\frac{\partial \bar{f}^i \sqrt{\overset{*}{a}}}{\partial \alpha^i} + \frac{\partial \bar{p}^{z*} \sqrt{\overset{*}{a}}}{\partial \overset{*}{z}} + \sqrt{\overset{*}{a}} \overset{*}{F} = 0, \tag{4.49}$$

where

$$\bar{f}^i = \overset{*}{\sigma}^{ik} \overset{*}{\bar{r}}_k + \overset{*}{\sigma}^{iz} \overset{*}{m} \quad (i,k = 1,2).$$

Here $\overset{*}{\sigma}^{ik}$, $\overset{*}{\sigma}^{iz}$ are the contravariant components of the stress tensor and the vector $\bar{p}^{z*}$ of external forces acting upon the faces $\overset{*}{z} = \pm 0.5h$ of the deformed shell, respectively, $\bar{F}$ is the vector of mass forces per unit volume of the deformed element, $\overset{*}{a} = \det(\overset{*}{a}_{ik})$. Integrating (4.49) and subsequently the vector product of (4.49) by $\overset{*}{m} \overset{*}{z}$ over the thickness of the shell, $z \in [z_1, z_2]$, we obtain

$$\frac{\partial \overset{*}{T}^{ik} \overset{*}{\bar{r}}_k \sqrt{\overset{*}{a}}}{\sqrt{\overset{*}{a} \partial \alpha^i}} + \bar{p}^{z*} \sqrt{\frac{\overset{*}{g}}{\overset{*}{a}}} \Bigg|_{z_1}^{z_2} + \int_{z_1}^{z_2} \sqrt{\frac{\overset{*}{g}}{\overset{*}{a}}} \overset{*}{F} d\overset{*}{z} = 0 \tag{4.50}$$

$$\left[\overset{*}{\bar{r}}_i, \overset{*}{T}^{ik} \overset{*}{\bar{r}}_k \right] = 0. \tag{4.51}$$

By introducing $\overset{*}{\bar{r}}_{ik}$ and $\overset{*}{m}_i$ given by (4.25) into (4.50) and equating to zero the coefficients for $\overset{*}{\bar{r}}_k$ and $\overset{*}{m}$, we find

$$\overset{*}{\nabla}_i \overset{*}{T}^{ik} + \sqrt{\frac{\overset{*}{g}}{\overset{*}{a}}} \overset{*}{\sigma}^{i3} (\delta_k^i - \overset{*}{z} \overset{*}{b}_{ik}) \Bigg|_{z_1}^{z_2} + \int_{z_1}^{z_2} \sqrt{\frac{\overset{*}{g}}{\overset{*}{a}}} \overset{*}{F}^i (\delta_k^i - \overset{*}{z} \overset{*}{b}_{ik}) d\overset{*}{z} = 0$$

$$\overset{*}{b}_{ik} \overset{*}{T}^{ik} + \sqrt{\frac{\overset{*}{g}}{\overset{*}{a}}} \overset{*}{\sigma}^{33} \Bigg|_{z_1}^{z_2} + \int_{z_1}^{z_2} \sqrt{\frac{\overset{*}{g}}{\overset{*}{a}}} \overset{*}{F}^3 d\overset{*}{z} = 0. \tag{4.52}$$

Here

$$\overset{*}{\nabla}_i \overset{*}{T}^{ik} = \frac{1}{\sqrt{\overset{*}{a}}} \frac{\partial \sqrt{\overset{*}{a}} \overset{*}{T}^{ik}}{\partial \alpha^i} + \overset{*}{\Gamma}_{ij}^k \overset{*}{T}^{ij} \quad (i,j = 1,2).$$

It follows immediately from (4.51) that the tensor $\overset{*}{T}{}^{ik}$ is symmetrical: $\overset{*}{T}{}^{12} = \overset{*}{T}{}^{21}$.

Although (4.50)–(4.52) are derived under the assumption $h = $ constant, they are also valid for shells of variable thickness $h = h(\alpha^1, \alpha^2)$.

Let the components σ^{ik} and in-plane forces T^{ik} be defined by

$$\sigma^{ik} = \overset{*}{\sigma}{}^{ik}\sqrt{\overset{*}{g}/g}, \quad T^{ik} = \overset{*}{T}{}^{ik}\sqrt{\overset{*}{a}/a}, \quad T^{ik} = \int_{z_1}^{z_2}\sqrt{\frac{g}{a}}\,\sigma^{ij}(\delta^i_k - \overset{*}{z}\,\overset{*}{b}_{ik})\mathrm{d}\overset{*}{z}\,.$$

Hence, the equations of equilibrium (4.52) in terms of the undeformed configuration take the form

$$\nabla_i T^{ik} + A^k_{ij} T^{ik} + \sqrt{\frac{g}{a}}\sigma^{i3}(\delta^i_k - \overset{*}{z}\,\overset{*}{b}_{ik})\Big|_{z_1}^{z_2} + \int_{z_1}^{z_2}\sqrt{\frac{g}{a}}F^i(\delta^i_k - \overset{*}{z}\,\overset{*}{b}_{ik})\mathrm{d}\overset{*}{z} = 0$$

$$\overset{*}{b}_{ik}T^{ik} + \sqrt{\frac{g}{a}}\sigma^{33}\Big|_{z_1}^{z_2} + \int_{z_1}^{z_2}\sqrt{\frac{g}{a}}F^3\mathrm{d}z = 0$$

$$T^{12} - T^{21} = 0, \tag{4.53}$$

where A^k_{ij} and $\nabla_i T^{ik}$ are calculated, with the use of (4.23), as

$$P_{n,ij} = \nabla_i\varepsilon_{nj} + \nabla_j\varepsilon_{ni} - \nabla_n\varepsilon_{ij}$$

$$\nabla_i T^{ik} = \frac{1}{\sqrt{a}}\frac{\partial\sqrt{a}T^{ik}}{\partial\alpha^i} + \Gamma^k_{ij}T^{ij}.$$

Exercises

In this chapter we have developed a general approach for describing geometry of the surface using orthogonal curvilinear coordinates. However, in the modeling of the pregnant uterus cylindrical coordinates become more practical. All exercises to follow are aimed at obtaining formulae of interest in cylindrical coordinates.

1. What are the advantages and limitations of the Kirchoff-Love hypotheses in the theory of shells?
2. Find the natural $\{\bar{r}_1, \bar{r}_2, \bar{m}\}$ and reciprocal $\{\bar{r}^1, \bar{r}^2, \bar{m}\}$ base on the surface S in cylindrical coordinates.
3. Find the unit base vectors $\bar{e}_i$ and $\bar{e}^i$ $(i = 1, 2)$ and the Lamé parameters of the surface.
4. Compute the covariant a_{ij} and contravariant a^{ij} $(i, j = 1, 2)$ components of the first metric tensor $\mathbf{A}$ of a surface.

5. Compute the covariant b_{ij} and contravariant b^{ij} $(i, j = 1, 2)$ components of the second metric tensor $\mathbf{B}$ of a surface.
6. Derive a formula for the first fundamental form of the surface in cylindrical coordinates.
7. Find the components of the second fundamental form $\overset{*}{b}_{ij}$ of $\overset{*}{S}$.
8. Verify (4.19).
9. Compute Γ^{i}_{jk} $(i, j, k = 1, 2)$.
10. Show that $\Gamma^{i}_{jk} = \Gamma^{i}_{kj}$.
11. Verify (4.14).
12. Verify (4.31).
13. Verify the Gauss-Codazzi equations given by (4.41).
14. Obtain the explicit form of the equation of equilibrium of a thin shell in terms of the undeformed configuration [*Hint*: consider (4.52) and (4.53)].

Chapter 5
Essentials of the Theory of Soft Shells

A mathematical theory is not to be considered complete until you have made it so clear that you can explain it to the first man whom you meet on the street.

D. Hilbert

5.1 Deformation of the Shell

A class of thin shells that:

1. $h/L \simeq 10^{-5}/10^{-2}$, L is the characteristic dimension of a shell
2. Possess zero-flexural rigidity
3. Do not withstand compression forces
4. Their actual configuration depends entirely on internal/external loads distributed over unit surface area
5. Undergo finite deformations
6. Their stress–strain states are fully described by in-plane membrane forces per unit length

is called soft shells. Results of this chapter are based on the work by Ridel and Gulin (1990) and the reader is advised to consult the original monograph for details.

Let the middle surface of the undeformed soft shell S be parameterized by curvilinear coordinates α^1, α^2 and $\forall M(\alpha^1, \alpha^2) \in S$ is described by a position vector $\bar{r}(\alpha^1, \alpha^2)$. As a result of the action of external and or internal loads, the shell will deform to attain a new configuration $\overset{*}{S}$. We assume that deformation is such that $\forall M(\alpha^1, \alpha^2) \to \overset{*}{M}(\alpha^1, \alpha^2)$ and is a homeomorphism. Thus, the inverse transformation exists. The transformation is defined analytically by

$$\overset{*}{\alpha^i} = \overset{*}{\alpha^i}(\alpha^1, \alpha^2), \tag{5.1}$$

R.N. Miftahof and H.G. Nam, *Biomechanics of the Gravid Human Uterus*,
DOI 10.1007/978-3-642-21473-8_5, © Springer-Verlag Berlin Heidelberg 2011

where (α^1, α^2) and $(\overset{*}{\alpha}{}^1, \overset{*}{\alpha}{}^2)$ are the coordinates on S and $\overset{*}{S}$, respectively. By assuming that (5.1) is continuously differentiable and $\det(\partial \overset{*}{\alpha}{}^i / \partial \alpha^k) \neq 0 \, (i, k = 1, 2)$, then the inverse transformation exists: $\alpha_i = \alpha_i \alpha^i = \alpha^i(\overset{*}{\alpha}{}^1, \overset{*}{\alpha}{}^2)$.

Let

$$C_i^k = \frac{\partial \overset{*}{\alpha}_k}{\partial \alpha_i}, \quad C = \det(C_i^k), \tag{5.2}$$

$$\overset{*}{C}_i^k = \frac{\partial \alpha_k}{\partial \overset{*}{\alpha}_i}, \quad \overset{*}{C} = \det(\overset{*}{C}_i^k). \tag{5.3}$$

By assuming that $C \neq 0$, $C^* \neq 0$, we have

$$C_i^k \overset{*}{C}_k^i = 1, \quad C_i^k \overset{*}{C}_j^k = 0 \quad (i \neq j),$$
$$C \overset{*}{C} = 1, \quad \overset{*}{C}_i^k = [C_k^i]/C. \tag{5.4}$$

Here, $[C_k^i]$ is the cofactor to the element C_k^i of the matrix (C_i^k).

Deformation of linear elements along the α^1, α^2 - coordinate lines is described by stretch ratios $\lambda_i (i = 1, 2)$ and elongations $e_{\alpha i}$

$$e_{\alpha i} = \frac{\overset{*}{ds_i} - ds_i}{ds_i} = \lambda_i - 1 = \frac{\sqrt{\overset{*}{a}_{ii}}}{\sqrt{a_{ii}}} - 1, \tag{5.5}$$

where $ds_i, \overset{*}{ds_i}$ - lengths of a line element between two infinitely close points on S and $\overset{*}{S}$, respectively, are given by

$$ds_i = \sqrt{a_{ii}} d\alpha^i, \quad \overset{*}{ds_i} = \sqrt{\overset{*}{a}_{ii}} d \overset{*}{\alpha}{}^i.$$

Changes in the angle, γ, between coordinate lines and the surface area, δs_Δ, are described by

$$\gamma = \overset{(0)}{\chi} - \overset{*}{\chi} = \overset{(0)}{\chi} - \cos^{-1} \frac{a_{12}}{\sqrt{a_{11} a_{22}}}. \tag{5.6}$$

$$\delta s_\Delta = \frac{\overset{*}{ds_\Delta}}{ds_\Delta} = \frac{\sqrt{\overset{*}{a}}}{\sqrt{a}} = \frac{\sqrt{\overset{*}{a}_{11} \overset{*}{a}_{22}} \sin \overset{*}{\chi}}{\sqrt{a_{11} a_{22}} \sin \overset{(0)}{\chi}} = \lambda_1 \lambda_2 \frac{\sin \overset{*}{\chi}}{\sin \overset{(0)}{\chi}}. \tag{5.7}$$

By making use of (5.2) and (5.3), for vectors $\bar{r}_i$, $\overset{*}{\bar{r}}_i$, tangent to coordinate lines on S and $\overset{*}{S}$, we have

$$\bar{r}_i = C_i^k \overset{*}{\bar{r}}_k, \, \overset{*}{\bar{r}}_i = \overset{*}{C}_i^k \bar{r}_k. \tag{5.8}$$

Hence, the unit vectors $\bar{e}_i \in S$, $\overset{*}{\bar{e}}_i \in \overset{*}{S}$ are found to be

$$\bar{e}_i = \frac{\bar{r}_i}{|\bar{r}_i|} = \frac{\bar{r}_i}{\sqrt{a_{ii}}} = C_i^k \left(\overset{*}{\bar{r}}_k \sqrt{\frac{\overset{*}{a}_{ii}}{\overset{*}{a}_{kk}}} \right) = \hat{C}_i^k \overset{*}{\bar{e}}_k,$$

$$\overset{*}{\bar{e}}_i = \overset{*}{\hat{C}}{}_i^k \bar{e}_k,$$

(5.9)

where the following notations are introduced:

$$\hat{C}_i^k = C_i^k \sqrt{\frac{\overset{*}{a}_{kk}}{a_{ii}}}, \qquad \overset{*}{\hat{C}}{}_i^k = \overset{*}{C}{}_i^k \sqrt{\frac{a_{kk}}{\overset{*}{a}_{ii}}},$$

(5.10)

With the help of (5.10), the scalar and vector products of unit vectors $\overset{*}{\bar{e}}_i$ and $\overset{*}{\bar{e}}_k$ are found to be

$$\overset{*}{\bar{e}}_i \cdot \overset{*}{\bar{e}}_k := \cos \overset{*}{\chi}_{i_k} = \bar{e}_j \cdot \bar{e}_n \hat{C}_i^j \hat{C}_k^n = \hat{C}_i^j \hat{C}_k^n \cos \overset{(0)}{\chi}_{jn},$$

$$\left[\overset{*}{\bar{e}}_i, \overset{*}{\bar{e}}_k \right] := \overset{*}{\bar{m}} \sin \overset{*}{\chi}_{i_k} = \bar{e}_j \times \bar{e}_n \hat{C}_i^j \hat{C}_k^n = \hat{C}_i^j \hat{C}_k^n \overset{*}{\bar{m}} \sin \overset{(0)}{\chi}_{jn}.$$

(5.11)

In just the same way, proceeding from the scalar and vector multiplication of $\bar{e}_i$ by $\bar{e}_k$, it can be shown that

$$\cos \overset{(0)}{\chi}_{ik} = \overset{*}{\hat{C}}{}_i^j \overset{*}{\hat{C}}{}_k^n \cos \overset{*}{\chi}_{jn},$$

$$\sin \overset{(0)}{\chi}_{ik} = \overset{*}{\hat{C}}{}_i^j \overset{*}{\hat{C}}{}_k^n \sin \overset{*}{\chi}_{jn}.$$

(5.12)

To calculate the coefficients C_i^k, $\overset{*}{\hat{C}}{}_i^k$, we proceed from geometric considerations. Let vectors $\bar{e}_i$, $\overset{*}{\bar{e}}_i$, at point $M(\alpha^1, \alpha^2) \in S$, be oriented as shown (Fig. 5.1). By decomposing $\overset{*}{\bar{e}}_i$ in the directions of $\bar{e}_k$, we have

$$\overset{*}{\hat{C}}{}_1^1 = \overline{MC}, \quad \overset{*}{\hat{C}}{}_1^2 = \overline{CD}, \quad \overset{*}{\hat{C}}{}_2^1 = -\overline{AB}, \quad \overset{*}{\hat{C}}{}_2^2 = \overline{MB}.$$

(5.13)

By solving ΔMCD and ΔMBA, we find

$$\overset{*}{\hat{C}}{}_1^1 = \frac{\sin(\overset{*}{\chi} - \overset{*}{\chi}_2)}{\sin \overset{*}{\chi}}, \quad \overset{*}{\hat{C}}{}_1^2 = \frac{\sin \overset{*}{\chi}_2}{\sin \overset{*}{\chi}},$$

$$\overset{*}{\hat{C}}{}_2^1 = -\frac{\sin(\overset{*}{\chi}_1 + \overset{*}{\chi}_2 - \overset{*}{\chi})}{\sin \overset{*}{\chi}}, \quad \overset{*}{\hat{C}}{}_2^2 = \frac{\sin(\overset{*}{\chi}_1 + \overset{*}{\chi}_2)}{\sin \overset{*}{\chi}},$$

$$\overset{*}{\hat{C}} = \det \overset{*}{\hat{C}}{}_i^k = \frac{\sin \overset{*}{\chi}_1}{\sin \overset{*}{\chi}}.$$

(5.14)

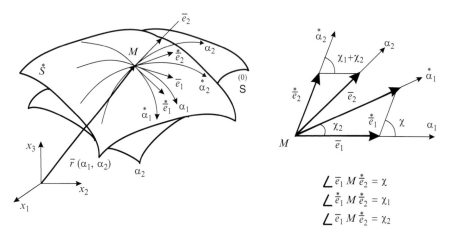

Fig. 5.1 Deformation of an element of the soft shell. Used with permission from Cambridge University Press

Similarly, by expanding unit vectors $\bar{e}_i$ along $\overset{*}{\bar{e}}_k$, we obtain

$$\hat{C}_1^1 = \frac{\sin(\overset{0}{\chi}_1 + \overset{0}{\chi}_2)}{\sin \overset{0}{\chi}_1}, \quad \hat{C}_1^2 = -\frac{\sin \overset{0}{\chi}_2}{\sin \overset{0}{\chi}_1},$$

$$\hat{C}_2^1 = \frac{\sin(\overset{0}{\chi}_1 + \overset{0}{\chi}_2 - \overset{0}{\chi})}{\sin \overset{0}{\chi}_1}, \quad \hat{C}_2^2 = \frac{\sin(\overset{0}{\chi} - \overset{0}{\chi}_2)}{\sin \overset{0}{\chi}_1},$$

$$\hat{C} = \det \hat{C}_i^k = \frac{\sin \overset{0}{\chi}}{\sin \overset{0}{\chi}_1}. \tag{5.15}$$

Note that the coefficients $C_i^k, \hat{C}_i^k$ are the functions of $\overset{0}{\chi}_i$ and $\overset{*}{\chi}_i$, while $\hat{C}_i^k, \overset{*}{\hat{C}}_i^k$ depend on $\bar{r}_i$ and $\overset{*}{\bar{r}}_i$ and the actual configuration of a shell.

Let the cut configuration of a soft shell $\overset{0}{S}$ be different from the undeformed configuration S. We introduce the coefficients of transformation $\bar{r}_i \in \overset{0}{S} \to \overset{*}{S}$ by

$$\overset{0}{\hat{C}}_i^k = C_i^k \sqrt{\overset{*}{a}_{kk} / \overset{0}{a}_{ii}}, \quad \overset{*}{\hat{C}}_i^k = C_i^k \sqrt{\overset{0}{a}_{kk} / \overset{*}{a}_{ii}}, \tag{5.16}$$

where $\overset{0}{a}_{ii}, \overset{*}{a}_{ii}$ are the components of the metric tensor $\mathbf{A}$ on $\overset{0}{S}$ and $\overset{*}{S}$, respectively. By eliminating $C_i^k, \overset{*}{C}_i^k$ from (5.10), for the coefficients of the cut and deformed surfaces, we obtain

$$\hat{C}_i^k = \overset{0}{\hat{C}}_i^k \frac{\lambda_k}{\lambda_i}, \quad \overset{*}{\hat{C}}_i^k = \overset{*}{\hat{C}}_i^k \frac{\lambda_k}{\overset{*}{\lambda}_i}. \tag{5.17}$$

Similarly, we introduce the coefficients

$$\overset{*}{\hat{C}}{}_i^k = \left[\hat{C}_k^i\right]/\hat{C}, \quad \overset{*}{\hat{\hat{C}}}{}_i^k = \left[\hat{\hat{C}}_k^i\right]/\hat{\hat{C}}, \tag{5.18}$$

where

$$\hat{C} = C\sqrt{\frac{\overset{*}{a}_{11}\,\overset{*}{a}_{22}}{\overset{0}{a}_{11}\,\overset{0}{a}_{22}}}, \quad \hat{\hat{C}} = C\sqrt{\frac{\overset{*}{a}_{11}\,\overset{*}{a}_{22}}{\overset{0}{a}_{11}\,\overset{0}{a}_{22}}}. \tag{5.19}$$

Finally, from (5.16) to (5.19), we get

$$\hat{C} = \hat{\hat{C}}\frac{\overset{*}{\lambda}_1\,\overset{*}{\lambda}_2}{\lambda_1\lambda_2}. \tag{5.20}$$

Let $\mathbf{E}$ be the tensor of deformation of S $(S = \overset{0}{S})$ given by

$$\mathbf{E} = \varepsilon_{ik}\bar{r}^i\bar{r}^k, \tag{5.21}$$

where

$$\varepsilon_{ik} = \frac{\overset{*}{a}_{ik} - a_{ik}}{2}. \tag{5.22}$$

By substituting (5.2) into (5.22) for ε_{ik}, we find

$$\varepsilon_{ik} = \frac{\left(\lambda_i\lambda_k \cos\overset{*}{\chi} - \cos\chi\right)\sqrt{a_{ii}a_{kk}}}{2}. \tag{5.23}$$

It is easy to show that the following relations hold:

$$\varepsilon_{ik} = \overset{*}{\varepsilon}_{jn} C_i^j C_k^n, \qquad \varepsilon^{ik} = \overset{*}{\varepsilon}^{jn} C_j^i C_n^k, \tag{5.24}$$

$$\overset{*}{\varepsilon}_{ik} = \varepsilon_{jn} \overset{*}{C}_i^j \overset{*}{C}_k^n, \qquad \overset{*}{\varepsilon}^{ik} = \varepsilon^{jn} C_j^i C_n^k.$$

In the theory of soft thin shells, stretch ratios and membrane forces per unit length of a differential element are preferred to traditional deformations and stresses per unit cross-sectional area of the shell. Thus, by dividing (5.23) by the surface area $\sqrt{a_{ii}a_{kk}}$ of an element, we get

$$\tilde{\varepsilon}_{ik} := \frac{\varepsilon_{ik}}{\sqrt{a_{ii}a_{kk}}} = \frac{\left(\lambda_i\lambda_k \cos\overset{*}{\chi} - \cos\chi\right)}{2}, \tag{5.25}$$

$\tilde{\varepsilon}_{ik}$ are called the physical components of **E**. Using (5.24), for $\overset{*}{\tilde{\varepsilon}}_{ik}$ in terms of the deformed configuration, we obtain

$$\overset{*}{\tilde{\varepsilon}}_{ik} = \tilde{\varepsilon}_{jn} \overset{*}{C}{}^{j}_{i} \overset{*}{C}{}^{n}_{k} \frac{\sqrt{a_{jj}a_{nn}}}{\sqrt{\overset{*}{a}_{jj}\,\overset{*}{a}_{nn}}}, \tag{5.26}$$

where the coefficients $\overset{*}{C}{}^{j}_{i}$ satisfy (5.2) and (5.3).

By making use of (5.25) in (5.3) and (5.6) for λ_i and γ, we find

$$\lambda_i = 1 + \varepsilon_i = \sqrt{1 + 2\tilde{\varepsilon}_{ii}},$$

$$\gamma = \overset{(0)}{\chi} - cos^{-1} \frac{2\tilde{\varepsilon}_{12} + \cos \overset{(0)}{\chi}}{\sqrt{(1 + 2\tilde{\varepsilon}_{11})}\sqrt{(1 + 2\tilde{\varepsilon}_{22})}}. \tag{5.27}$$

By substituting $\overset{\hat{}}{\hat{C}}{}^{k}_{i}, \overset{*}{\hat{C}}{}^{k}_{i}$ given by (5.16) in (5.26), we have

$$\overset{*}{\tilde{\varepsilon}}_{ik} = \tilde{\varepsilon}_{jn} \overset{\hat{}}{\hat{C}}{}^{j}_{i} \overset{*}{\hat{C}}{}^{n}_{k} \quad \tilde{\varepsilon}_{ik} = \tilde{\varepsilon}^{m}{}_{jn}\overset{\hat{}}{\hat{C}}{}^{j}_{i}\overset{\hat{}}{\hat{C}}{}^{n}_{k}. \tag{5.28}$$

Finally, formulas for $\tilde{\varepsilon}_{ik}$ in terms of $\overset{0}{S}$ - configuration of the soft shell take the form

$$\overset{*}{\tilde{\varepsilon}}_{11} = \frac{1}{\sin^2 \overset{0}{\chi}}\left[\tilde{\varepsilon}_{11}\sin^2(\overset{0}{\chi} - \overset{0}{\chi}_2) + \tilde{\varepsilon}_{22}\sin^2 \overset{0}{\chi}_2 + 2\tilde{\varepsilon}_{12}\sin(\overset{0}{\chi} - \overset{0}{\chi}_2)\sin \overset{0}{\chi}_2\right],$$

$$\overset{*}{\tilde{\varepsilon}}_{12} = \frac{1}{\sin^2 \overset{0}{\chi}}\left[-\tilde{\varepsilon}_{11}\sin(\overset{0}{\chi} - \overset{0}{\chi}_2)\sin(\overset{0}{\chi}_1 + \overset{0}{\chi}_2 - \overset{0}{\chi}) + \tilde{\varepsilon}_{22}\sin \overset{0}{\chi}_2 \sin(\overset{0}{\chi}_1 + \overset{0}{\chi}_2)\right.$$
$$\left. + \tilde{\varepsilon}_{12}\left(\cos(\overset{0}{\chi}_1 + 2\overset{0}{\chi}_2 - \overset{0}{\chi}) - \cos \overset{0}{\chi} \cos \overset{0}{\chi}_1\right)\right],$$

$$\overset{*}{\tilde{\varepsilon}}_{22} = \frac{1}{\sin^2 \overset{0}{\chi}}\left[\tilde{\varepsilon}_{11}\sin^2(\overset{0}{\chi}_1 + \overset{0}{\chi}_2 - \overset{0}{\chi}) + \tilde{\varepsilon}_{22}\sin^2(\overset{0}{\chi}_1 + \overset{0}{\chi}_2)\right.$$
$$\left. - 2\tilde{\varepsilon}_{12}\sin(\overset{0}{\chi}_1 + \overset{0}{\chi}_2 - \overset{0}{\chi})\sin(\overset{0}{\chi}_1 + \overset{0}{\chi}_2)\right]. \tag{5.29}$$

With the help of (5.25) in (5.28), the physical components can also be expressed in terms of stretch ratios and shear angles as

$$\overset{*}{\lambda}_i \overset{*}{\lambda}_k \cos \overset{*}{\chi}_{ik} - \cos \overset{0}{\chi}_{ik} = (\lambda_j\lambda_n \cos \chi_{jn} - \cos \overset{0}{\chi}_{jn}) \overset{\hat{}}{\hat{C}}{}^{j}_{i} \overset{*}{\hat{C}}{}^{n}_{k}. \tag{5.30}$$

Further, on use of (5.11) and (5.14), (5.30) in terms of $\overset{0}{S}$ - configuration takes the form

$$\overset{*}{\lambda}_i \overset{*}{\lambda}_k \cos \overset{*}{\chi}_{ik} = \lambda_j\lambda_n \overset{\hat{}}{\hat{C}}{}^{j}_{i} \overset{*}{\hat{C}}{}^{n}_{k} \cos \overset{0}{\chi}_{jn}, \tag{5.31}$$

or in expanded form

$$
\overset{*}{\lambda}_1 = \frac{1}{\sin^2 \overset{0}{\chi}}\left[\lambda_1^2\sin^2(\overset{0}{\chi}-\overset{0}{\chi}_2) + \lambda_2^2\sin^2\overset{0}{\chi}_2+2\lambda_1\lambda_2 \sin(\overset{0}{\chi}-\gamma)\sin(\overset{0}{\chi}-\overset{0}{\chi}_2)\sin\overset{0}{\chi}_2\right]^{1/2},
$$

$$
\overset{*}{\gamma} = \overset{0}{\chi}_1 - \cos^{-1}\left[\left(-\lambda_1^2\sin(\overset{0}{\chi}-\overset{0}{\chi}_2)\sin(\overset{0}{\chi}_1+\overset{0}{\chi}_2-\overset{0}{\chi})+\lambda_2^2\sin\overset{0}{\chi}_2\sin(\overset{0}{\chi}_1+\overset{0}{\chi}_2)\right.\right.
$$

$$
\left.\left.+ \lambda_1\lambda_2(\cos(\widehat{\delta}+2\overset{0}{\chi}_2-\overset{0}{\chi}) - \cos\overset{0}{\chi}\cos\overset{0}{\chi}_1)\cos(\overset{0}{\chi}-\gamma)\right)\left(\overset{*}{\lambda}_1\overset{*}{\lambda}_2\sin^2\overset{0}{\chi}\right)^{-1}\right]
$$

$$
\overset{*}{\lambda}_2 = \frac{1}{\sin^2 \overset{0}{\chi}}\left[\lambda_1^2\sin^2(\overset{0}{\chi}_1+\overset{0}{\chi}_2-\overset{0}{\chi}) + \lambda_2^2\sin^2(\overset{0}{\chi}_1+\overset{0}{\chi}_2)\right.
$$

$$
\left. - 2\lambda_1\lambda_2 \cos(\overset{0}{\chi}-\Delta\overset{*}{\chi})\sin(\overset{0}{\chi}_1+\overset{0}{\chi}_2-\overset{0}{\chi})\sin(\overset{0}{\chi}_1+\overset{0}{\chi}_2)\right]^{1/2}.
$$

$$(5.32)$$

Formulas (5.31) and (5.32) are preferred in practical applications particularly when dealing with finite deformations of shells.

5.2 Principal Deformations

At any point $\overset{*}{M} \in \overset{*}{S}$, there exist two mutually orthogonal directions that remain orthogonal during deformation and along which the components of $\mathbf{E}$ attain the maximum and minimum value. They are called the principal directions.

To find the orientation of the principal axes, we proceed as follows. Let $\overset{(0)}{\varphi}, \overset{*}{\varphi}$ be the angles of the direction away from the base vectors $\bar{e}_1 \in \overset{(0)}{S}, \overset{*}{\bar{e}}_1 \in \overset{*}{S}$, respectively. We assume that the cut and undeformed configurations are indistinguisable $\overset{0}{S} = S$. Then, by setting $\overset{0}{\chi}_2 = \overset{0}{\varphi}$ in the first equation of (5.29), we have

$$
\tilde{\varepsilon}_{11}^*\sin^2 \overset{0}{\chi} = \tilde{\varepsilon}_{11}\sin^2(\overset{0}{\chi}-\overset{0}{\varphi}) + 2\tilde{\varepsilon}_{12}\sin(\overset{0}{\chi}-\overset{0}{\varphi})\sin\overset{0}{\varphi}+\tilde{\varepsilon}_{22}\sin^2\overset{0}{\varphi}. \tag{5.33}
$$

After simple rearrangements, it can be written in the form

$$
\varepsilon = a_0 + b_0 \cos 2\overset{0}{\varphi}+c_0 \sin 2\overset{0}{\varphi}, \tag{5.34}
$$

where

$$
a_0 = \frac{1}{\sin^2 \overset{0}{\chi}}\left[\frac{1}{2}(\tilde{\varepsilon}_{11}+\tilde{\varepsilon}_{22}) - \tilde{\varepsilon}_{12}\cos\overset{0}{\chi}\right],
$$

$$
b_0 = \frac{1}{\sin^2 \overset{0}{\chi}}\left[\frac{1}{2}(\tilde{\varepsilon}_{11}-\tilde{\varepsilon}_{22}) + \left(\tilde{\varepsilon}_{12}-\tilde{\varepsilon}_{11}\cos\overset{0}{\chi}\right)\Big/\cos\overset{0}{\chi}\right], \tag{5.35}
$$

$$
c_0 = \frac{1}{\sin^2 \overset{0}{\chi}}\left(\tilde{\varepsilon}_{12}-\tilde{\varepsilon}_{11}\cos\overset{0}{\chi}\right).
$$

By differentiating (5.34) with respect to $\overset{0}{\varphi}$ and equating the resultant equation to zero, for the principal axes on the surface $\overset{(0)}{S}$, we find ($b_0 \neq 0$)

$$\tan 2 \overset{0}{\varphi} = \frac{c_0}{b_0} = \frac{2 \left(\tilde{\varepsilon}_{11} \cos \overset{0}{\chi} - \tilde{\varepsilon}_{12} \right) \sin \overset{0}{\chi}}{\tilde{\varepsilon}_{11} \cos 2 \overset{0}{\chi} - 2 \tilde{\varepsilon}_{12} \cos \overset{0}{\chi} + \tilde{\varepsilon}_{22}}. \tag{5.36}$$

By substituting (5.36) into (5.34), we obtain the principal physical components $\varepsilon_1, \varepsilon_2$ of $\mathbf{E}$

$$\varepsilon_{1,2} = a_0^2 \pm \sqrt{b_0^2 + c_0^2} = \frac{(\tilde{\varepsilon}_{11} + \tilde{\varepsilon}_{22}) - 2\tilde{\varepsilon}_{12} \cos \overset{0}{\chi}}{2 \sin^2 \overset{0}{\chi}}$$

$$\pm \frac{1}{\sin^2 \overset{0}{\chi}} \sqrt{\frac{(\tilde{\varepsilon}_{11} - \tilde{\varepsilon}_{22})^2}{4} + \tilde{\varepsilon}_{12}^2 + \tilde{\varepsilon}_{11}\tilde{\varepsilon}_{22}\cos^2 \overset{0}{\chi} - \tilde{\varepsilon}_{12}(\tilde{\varepsilon}_{11} + \tilde{\varepsilon}_{22}) \cos \overset{0}{\chi}}. \tag{5.37}$$

Henceforth, we assume that $\max \varepsilon_1$ is achieved in the direction of the principal axis defined by the angle $\overset{0}{\varphi} = \overset{0}{\varphi}$, and $\min \varepsilon_2$ along the axis, defined by the angle $\overset{0}{\varphi_2} = \overset{0}{\varphi} + \pi/2$. Since for the principal directions $\overset{0}{\chi} \equiv \pi/2$, from the second (5.29), we find

$$\overset{*}{\tilde{\varepsilon}}_{12} = -b_0 \sin 2 \overset{0}{\varphi} + c_0 \cos 2 \overset{0}{\varphi}. \tag{5.38}$$

By dividing both sides of (5.39) by $c_0 \cos 2 \overset{0}{\varphi}$ and using (5.37), we find $\overset{*}{\tilde{\varepsilon}}_{12} = 0$. Thus, there exist indeed two mutually orthogonal directions at $\forall M(\alpha^i) \in \overset{(0)}{S}$ that remain orthogonal throughout deformation. Similar result, i.e., $\overset{*}{\gamma} = 0$, can be obtained from (5.13) by setting $\overset{0}{\chi} = \pi/2$.

By substituting (5.25) in (5.36), (5.37) for the orientation of the principal axes on $\overset{0}{S}$ and the principal stretch ratios, we obtain

$$\tan 2 \overset{0}{\underset{1}{\varphi}} = \frac{2 \left[\lambda_1 \lambda_2 \cos(\overset{0}{\chi} - \overset{*}{\gamma}) - \lambda_1^2 \cos \overset{0}{\chi} \right] \sin \overset{0}{\chi}}{\lambda_1^2 - \lambda_2^2 + 2 \left[\lambda_1 \lambda_2 \cos(\overset{0}{\chi} - \overset{*}{\gamma}) - \lambda_1^2 \cos \overset{0}{\chi} \right] \cos \overset{0}{\chi}},$$

$$\overset{0}{\underset{2}{\varphi}} = \overset{0}{\underset{1}{\varphi}} + \pi/2, \tag{5.39}$$

$$\Lambda_{1,2}^2 = \frac{1}{\sin \overset{0}{\chi}} \left[(\lambda_1^2 + \lambda_2^2)/2 - \lambda_1 \lambda_2 \cos(\overset{0}{\chi} - \overset{*}{\gamma}) \cos \overset{0}{\chi} \right.$$

$$\pm \left((\lambda_1^2 + \lambda_2^2)^2/4 + \lambda_1^2 \lambda_2^2 \cos(2\overset{0}{\chi} - \overset{*}{\gamma}) \cos \overset{*}{\gamma} \right.$$

$$\left. - \lambda_1 \lambda_2 (\lambda_1^2 + \lambda_2^2) \cos(\overset{0}{\chi} - \overset{*}{\gamma}) \cos \overset{0}{\chi} \right)^{1/2} \Big]^{1/2}. \tag{5.40}$$

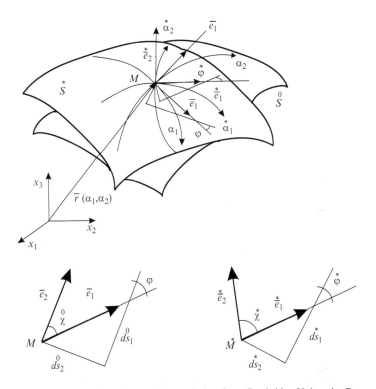

Fig. 5.2 Principal deformations. Used with permission from Cambridge University Press

To find the orientation of the principal axes on the deformed surface $\overset{*}{S}$, consider a triangular element on S bounded by the two principal axes and the α_1 - coordinate line (Fig. 5.2). Geometric analysis leads to the following obvious equalities:

$$\cos \overset{*}{\varphi}_1 = \frac{\mathrm{d}\overset{*}{s}_1}{\mathrm{d}s_1} = \frac{\Lambda_1}{\lambda_1} \cos \overset{(0)}{\varphi}_1, \quad \sin \overset{*}{\varphi}_1 = \frac{\mathrm{d}\overset{*}{s}_2}{\mathrm{d}s_1} = \frac{\Lambda_2}{\lambda_1} \sin \overset{(0)}{\varphi}_1$$

$$\tan \overset{*}{\varphi}_1 = \frac{\mathrm{d}\overset{*}{s}_2}{\mathrm{d}\overset{*}{s}_1} = \frac{\Lambda_2}{\Lambda_1} \tan \overset{(0)}{\varphi}_1. \tag{5.41}$$

On use of (5.39) and (5.40) from the above, we find the angles for the principal axes $\overset{*}{\varphi}$ and $\overset{*}{\varphi}_2 = \overset{*}{\varphi} + \pi/2$.

Finally, by substituting (5.7) and (5.25) into expressions for the first and second invariants of the tensor of deformation **E** defined by

$$I_1^{(E)} = \varepsilon_1 + \varepsilon_2 = \overset{*}{\tilde{\varepsilon}}_{11} + \overset{*}{\tilde{\varepsilon}}_{22} = \frac{1}{\sin^2 \overset{0}{\chi}} \left[\tilde{\varepsilon}_{11} - 2\tilde{\varepsilon}_{12} \cos \overset{0}{\chi} + \tilde{\varepsilon}_{22} \right],$$

$$I_2^{(E)} = \varepsilon_1 \varepsilon_2 = \overset{*}{\tilde{\varepsilon}}_{11} \overset{*}{\tilde{\varepsilon}}_{22} - (\overset{*}{\tilde{\varepsilon}}_{12})^2 = \frac{1}{\sin^2 \overset{0}{\chi}} \left[\bar{\varepsilon}_{11} \bar{\varepsilon}_{22} - (\bar{\varepsilon}_{12})^2 \right]. \tag{5.42}$$

for the principal stretch ratios and the shear angle, we get

$$\Lambda_1^2 + \Lambda_2^2 = (\overset{*}{\lambda}_1)^2 + (\overset{*}{\lambda}_2)^2 = \frac{1}{\sin^2 \overset{0}{\chi}} \left(\lambda_1^2 + \lambda_2^2 - 2\lambda_1\lambda_2 \cos(\overset{0}{\chi} - \gamma) \cos \overset{0}{\chi} \right),$$

$$\Lambda_1\Lambda_2 = \sqrt{1 + 2I_1^{(E)} + 4I_2^{(E)}} = \overset{*}{\lambda}_1 \overset{*}{\lambda}_2 \cos\gamma = \frac{\lambda_1\lambda_2 \sin(\overset{0}{\chi} - \gamma)}{\sin \overset{0}{\chi}}.$$

(5.43)

The last equation is also used to calculate the change of the surface area of S.

5.3 Membrane Forces

The stress state of a differential element of the soft shell is described entirely by in-plane tangent $T_{ii}(T^{ii})$ and shear $T_{ik}(T^{ik})(i \neq k)$ forces per unit length of the element. To study the equilibrium of the shell, we proceed from consideration of triangular

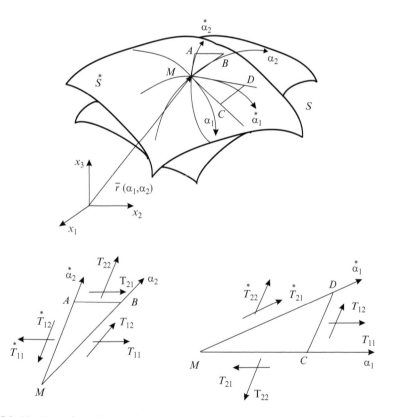

Fig. 5.3 Membrane forces in the soft shell. Used with permission from Cambridge University Press

elements ΔMAB and ΔMCD on $\overset{*}{S}$ (Fig. 5.3). Analysis of force distribution in the elements yields

$$-\overline{MA}\,\overset{*}{T}{}^{1k}\underset{k}{\bar{e}} + \overline{MB}T^{1i}\,\bar{e}_i + \overline{AB}T^{2i}\,\bar{e}_i = 0,$$

$$\overline{MD}\,\overset{*}{T}{}^{2k}\underset{k}{\bar{e}} + \overline{CD}T^{1i}\,\bar{e}_i - \overline{MC}T^{2i}\,\bar{e}_i = 0, \tag{5.44}$$

where $\overset{*}{\hat{C}}{}^1_1 = \overline{MC}$, $\overset{*}{\hat{C}}{}^2_1 = \overline{CD}$, $\overset{*}{\hat{C}}{}^1_2 = -\overline{AB}$, $\overset{*}{\hat{C}}{}^2_2 = \overline{MB}$ (5.13). The scalar product of (5.44) and $\bar{e}^k$ yields

$$\overset{*}{T}{}^{1k} = T^{1i}\hat{C}^k_i\,\overset{*}{\hat{C}}{}^2_2 - T^{2i}\hat{C}^k_i\,\overset{*}{\hat{C}}{}^1_2,$$

$$\overset{*}{T}{}^{2k} = -T^{1i}\hat{C}^k_i\,\overset{*}{\hat{C}}{}^2_1 + T^{2i}\hat{C}^k_i\,\overset{*}{\hat{C}}{}^1_1.$$

where use is made of (5.9). By substituting $\overset{*}{\hat{C}}{}^k_i$ given by (5.18) for $\hat{C}^k_i$, we find

$$\overset{*}{T}{}^{ik} = \frac{1}{C}T^{jn}\hat{C}^i_j\hat{C}^k_n. \tag{5.45}$$

On use of (5.15), the components of the membrane forces are found to be

$$\overset{*}{T}{}^{11} = \frac{1}{\sin\overset{0}{\chi}\sin\overset{0}{\chi}_1}\Big\{ T^{11}\sin^2(\overset{0}{\chi}_1 + \overset{0}{\chi}_2) + T^{22}\sin^2(\overset{0}{\chi}_1 + \overset{0}{\chi}_2 - \overset{0}{\chi})$$

$$+ 2T^{12}\sin(\overset{0}{\chi}_1 + \overset{0}{\chi}_2 - \overset{0}{\chi})\sin(\overset{0}{\chi}_1 + \overset{0}{\chi}_2)\Big\},$$

$$\overset{*}{T}{}^{12} = \frac{1}{\sin\overset{0}{\chi}\sin\overset{0}{\chi}_1}\Big\{ -T^{11}\sin\overset{0}{\chi}_2\sin(\overset{0}{\chi}_1 + \overset{0}{\chi}_2) + T^{22}\sin(\overset{0}{\chi}_1 + \overset{0}{\chi}_2 - \overset{0}{\chi})\sin(\overset{0}{\chi} - \overset{0}{\chi}_2)$$

$$+ T^{12}\Big[\cos(\overset{0}{\chi}_1 + 2\overset{0}{\chi}_2 - \overset{0}{\chi}) - \cos\overset{0}{\chi}\cos\overset{0}{\chi}_1\Big]\Big\},$$

$$\overset{*}{T}{}^{22} = \frac{1}{\sin\overset{0}{\chi}\sin\overset{0}{\chi}_1}\Big\{ T^{11}\sin^2\overset{0}{\chi}_2 - 2T^{12}\sin(\overset{0}{\chi} - \overset{0}{\chi}_2)\sin\overset{0}{\chi}_2 + T^{22}\sin^2(\overset{0}{\chi} - \overset{0}{\chi}_2)\Big\}. \tag{5.46}$$

On introducing the tensor of membrane forces $\mathbf{T}$

$$\mathbf{T} = T^{ik}\bar{r}_i\bar{r}_k = \frac{1}{\sin\overset{*}{\chi}}\tilde{T}^{ik}\bar{e}_i\bar{e}_k, \tag{5.47}$$

where $\tilde{T}^{ik}$ are the physical components of $\mathbf{T}$, and using (5.9), $\tilde{T}^{ik}$ can be expressed in terms of $\overset{*}{\tilde{T}}{}^{ik}$ as

$$\mathbf{T} = \frac{1}{\sin\overset{*}{\chi}}\tilde{T}^{ik}\bar{e}_i\bar{e}_k = \frac{1}{\sin\overset{*}{\chi}_1}\overset{*}{\tilde{T}}{}^{jn}\underset{ik}{\bar{e}\bar{e}} = \frac{1}{\sin\overset{*}{\chi}_1}\overset{*}{\tilde{T}}{}^{jn}\bar{e}_i\bar{e}_k\overset{*}{\hat{C}}{}^i_j\,\overset{*}{\hat{C}}{}^k_n.$$

Further, by making use of (5.15), we obtain

$$\mathbf{T} = \frac{1}{\overset{*}{C}} \overset{*}{\tilde{T}}{}^{jn} \overset{*}{\hat{C}}{}^i_j \overset{*}{\hat{C}}{}^k_n. \tag{5.48}$$

By substituting $\overset{*}{\hat{C}}{}^i_j$ given by (5.14), we get

$$\tilde{T}^{11} = \frac{1}{\sin\overset{*}{\chi}\sin\overset{*}{\chi}_1}\left\{\overset{*}{\tilde{T}}{}^{11}\sin^2(\overset{*}{\chi}-\overset{*}{\chi}_2)+\overset{*}{\tilde{T}}{}^{22}\sin^2(\overset{*}{\chi}_1+\overset{*}{\chi}_2-\overset{*}{\chi})\right.$$
$$\left.-2\overset{*}{\tilde{T}}{}^{12}\sin(\overset{*}{\chi}_1+\overset{*}{\chi}_2-\overset{*}{\chi})\sin(\overset{*}{\chi}-\overset{*}{\chi}_2)\right\},$$

$$\tilde{T}^{12} = \left\{\overset{*}{\tilde{T}}{}^{11}\sin\overset{*}{\chi}_2\sin(\overset{*}{\chi}-\overset{*}{\chi}_2)+\overset{*}{\tilde{T}}{}^{22}\sin(\overset{*}{\chi}_1+\overset{*}{\chi}_2-\overset{*}{\chi})\sin(\overset{*}{\chi}_1+\overset{*}{\chi}_2)\right.$$
$$\left.+\overset{*}{\tilde{T}}{}^{12}\left[\cos(\overset{*}{\chi}_1+2\overset{*}{\chi}_2-\chi)-\cos\overset{*}{\chi}\cos\overset{*}{\chi}_1\right]\right\}\bigg/\sin\overset{*}{\chi}\sin\overset{*}{\chi}_1,$$

$$\tilde{T}^{22} = \frac{1}{\sin\overset{*}{\chi}\sin\overset{*}{\chi}_1}\left\{\overset{*}{\tilde{T}}{}^{11}\sin\overset{*}{\chi}_2-2\overset{*}{\tilde{T}}{}^{12}\sin(\overset{*}{\chi}_1+\overset{*}{\chi}_2)\sin\overset{*}{\chi}_2+\overset{*}{\tilde{T}}{}^{12}\sin^2(\overset{*}{\chi}_1+\overset{*}{\chi}_2)\right\}.$$

$$\tag{5.49}$$

Using (5.17) after simple rearrangements, (5.48) takes the form

$$\frac{\lambda_1\lambda_2}{\lambda_i\lambda_k}\tilde{T}^{ik} = \frac{1}{\overset{\hat{*}}{C}}\frac{\overset{*}{\lambda}_1\overset{*}{\lambda}_2}{\overset{*}{\lambda}_j\overset{*}{\lambda}_n}\overset{*}{\tilde{T}}{}^{jn}\overset{\hat{*}}{\hat{C}}{}^{*j}_j\overset{\hat{*}}{\hat{C}}{}^{*k}_n. \tag{5.50}$$

Formulas (5.50) are preferred to (5.48) in applications. First, the coefficients $\overset{\hat{*}}{\hat{C}}{}^{*k}_n$ are used in calculations of both deformations and membrane forces. Second, $\overset{\hat{*}}{\hat{C}}{}^{*k}_n$ depend only on parameterization of the initial configuration of the shell. Therefore, once calculated they can be used throughout.

5.4 Principal Membrane Forces

As in the case of principal deformations at any point $\overset{*}{M} \in \overset{*}{S}$, there exist two mutually orthogonal directions that remain orthogonal throughout deformation and along which $\mathbf{T}$ attains the extreme values. They are called the principal directions and the principal membrane forces, respectively.

By assuming that the coordinates $\overset{*}{\alpha}_i \in \overset{*}{S}$ and $\alpha_i \in S$ are related by the angle $\overset{*}{\psi}$, and then by setting $\overset{*}{\chi}_1 = \pi/2$ and $\overset{*}{\chi}_2 = \psi$ in (5.46), we find

$$\overset{*}{T}{}^{11} = \frac{1}{\sin \overset{*}{\chi}} \left\{ T^{11}\cos^2 \overset{*}{\psi} + T^{22}\cos^2(\overset{*}{\chi} - \overset{*}{\psi}) + 2T^{12} \cos \overset{*}{\psi} \cos(\overset{*}{\chi} - \overset{*}{\psi}) \right\},$$

$$\overset{*}{T}{}^{12} = \frac{1}{\sin \overset{*}{\chi}} \left\{ -T^{11} \cos \overset{*}{\psi} \sin \overset{*}{\psi} + T^{12} \sin(\overset{*}{\chi} - 2\overset{*}{\psi}) + T^{22} \cos(\overset{*}{\chi} - \overset{*}{\psi}) \sin(\overset{*}{\chi} - \overset{*}{\psi}) \right\},$$

$$\overset{*}{T}{}^{22} = \frac{1}{\sin \overset{*}{\chi}} \left\{ T^{11}\sin^2 \overset{*}{\psi} - 2T^{12} \sin(\overset{*}{\chi} - \overset{*}{\psi}) \sin \overset{*}{\psi} + T^{22}\sin^2(\overset{*}{\chi} - \overset{*}{\psi}) \right\}.$$

$$(5.51)$$

Equation (5.51) can be written in the form

$$\overset{*}{T}{}^{11} = a_1 + b_1 \cos 2\overset{*}{\psi} + c_1 \sin 2\overset{*}{\psi},$$
$$\overset{*}{T}{}^{12} = -b_1 \sin 2\overset{*}{\psi} + c_1 \cos 2\overset{*}{\psi}, \qquad (5.52)$$
$$\overset{*}{T}{}^{22} = a_1 - b_1 \cos 2\overset{*}{\psi} - c_1 \sin 2\overset{*}{\psi},$$

where the following notations are introduced:

$$a_1 = \frac{1}{\sin \overset{*}{\chi}} \left[\frac{1}{2}(T^{11} + T^{22}) - T^{12} \cos \overset{*}{\chi} \right],$$

$$b_1 = \frac{1}{\sin \overset{*}{\chi}} \left[\frac{1}{2}(T^{11} + T^{22}) + \left(T^{12} + T^{22} \cos \overset{*}{\chi} \right) \cos \overset{*}{\chi} \right],$$

$$c_1 = T^{12} + T^{22} \cos \overset{*}{\chi}.$$

By differentiating $\overset{*}{T}{}^{ii}$ with respect to $\overset{*}{\psi}$ and equating the result to zero, we obtain

$$\tan 2\overset{*}{\psi} = \frac{c_1}{b_1} = \frac{2\left(T^{12} + T^{22} \cos \overset{*}{\chi} \right) \sin \overset{*}{\chi}}{T^{11} + 2T^{12} \cos \overset{*}{\chi} + T^{22} \cos 2 \overset{*}{\chi}}. \qquad (5.53)$$

By solving the above for $\overset{*}{\psi}$ for the directional angles of the principal axes, we get

$$\tan 2\overset{*}{\psi}_1 = \frac{2\left(T^{12} + T^{22} \cos \overset{*}{\chi} \right) \sin \overset{*}{\chi}}{T^{11} + 2T^{12} \cos \overset{*}{\chi} + T^{22} \cos 2 \overset{*}{\chi}},$$

$$\overset{*}{\psi}_2 = \overset{*}{\psi}_1 + \pi/2. \qquad (5.54)$$

By substituting (5.53) into (5.54), we have

$$\overset{*}{T}{}^{11} = a_1 + \sqrt{b_1^2 + c_1^2}, \qquad \overset{*}{T}{}^{12} = 0, \qquad \overset{*}{T}{}^{22} = a_1 - \sqrt{b_1^2 + c_1^2}.$$

From (5.52) the principal membrane forces T_1, T_2 are found to be

$$T_{1,2} = a_1{}^2 \pm \sqrt{b_1{}^2 + c_1{}^2} = \frac{1}{\sin \overset{*}{\chi}} \left\{ \frac{(T^{11} + T^{22})}{2} + T^{12} \cos \overset{*}{\chi} \right.$$

$$\left. \pm \sqrt{1/4(T^{11} - T^{22})^2 + (T^{12})^2 + T^{12}(T^{11} + T^{22}) \cos \overset{*}{\chi} + T^{11} T^{22} \cos^2 \overset{*}{\chi}} \right\}.$$

$$(5.55)$$

Thus, at each point of the surface of the soft shell there are two mutually orthogonal directions that remain orthogonal throughout deformation. Henceforth, we assume that $T_1 \geq T_2$; i.e., the maximum stress is in the direction of the principal axis defined by the angle $\overset{*}{\psi}_1$, and the minimum – by the angle $\overset{*}{\psi}_2$.

Analogously to the invariants of the tensor of deformation described by (5.43), we introduce the first and second invariants of T

$$I^{(T)}{}_1 = T_1 + T_2 = \overset{*}{T}{}^{11} + \overset{*}{T}{}^{22} = \frac{1}{\sin \overset{*}{\chi}} \left(T^{11} + T^{22} + 2T^{12} \cos \overset{*}{\chi} \right),$$

$$I^{(T)}{}_2 = T_1 T_2 = \overset{*}{T}{}^{11} \overset{*}{T}{}^{22} - (\overset{*}{T}{}^{12})^2 = T^{11} T^{22} - (T^{12})^2.$$

$$(5.56)$$

5.5 Equations of Motion in General Curvilinear Coordinates

Let $\Delta \overset{(*)}{\sigma}$ and $\Delta \overset{(*)}{m}$ be the surface area and mass of a differential element of the soft shell in undeformed and deformed configurations. Position of a point $\overset{*}{M} \in \overset{*}{S}$ at any moment of time t is given by vector $\bar{r}(\alpha^1, \alpha^2, t)$. Densities of the material in undeformed, ρ, and deformed, $\overset{*}{\rho}$, states are defined by

$$\rho = \lim_{\Delta\sigma \to 0} \frac{\Delta m}{\Delta \sigma} = \frac{dm}{d\sigma} \quad \text{and} \quad \overset{*}{\rho} = \lim_{\Delta\overset{*}{\sigma} \to 0} \frac{\Delta \overset{*}{m}}{\Delta \overset{*}{\sigma}} = \frac{d\overset{*}{m}}{d\overset{*}{\sigma}}, \qquad (5.57)$$

where

$$d\sigma = \sqrt{a} d\alpha^1 d\alpha^2, \quad d\overset{*}{\sigma} = \sqrt{\overset{*}{a}} d\alpha^1 d\alpha^2,$$

By applying the law of conservation of the mass to (5.57), we find

$$dm = d\overset{*}{m} = \overset{*}{\rho} d\overset{*}{\sigma} = \overset{*}{\rho} \sqrt{\overset{*}{a}} d\alpha^1 d\alpha^2 = \rho \sqrt{a} d\alpha^1 d\alpha^2.$$

It follows that

$$\overset{*}{\rho} = \rho \sqrt{\frac{a}{\overset{*}{a}}}. \tag{5.58}$$

Let $\bar{p}_s(\alpha_1, \alpha_2, t)$ be the resultant of the external, $\bar{p}_{(+)}(\alpha_1, \alpha_2, t)$, and internal, $\bar{p}_{(-)}(\alpha_1, \alpha_2, t)$, forces distributed over the outer and inner surfaces of the shell

$$\bar{p}_s(\alpha_1, \alpha_2, t) = \bar{p}_{(+)}(\alpha_1, \alpha_2, t) + \bar{p}_{(-)}(\alpha_1, \alpha_2, t).$$

The density of the resultant force per unit area of a deformed element $\bar{p}_s$ is defined by

$$\bar{p}(\alpha_1, \alpha_2, t) = \lim_{\Delta\sigma^* \to 0} \frac{\bar{p}_s}{\Delta\overset{*}{\sigma}}. \tag{5.59}$$

Similarly, we introduce the density of the mass force $\bar{F}(\alpha^1, \alpha^2, t)$ by

$$\bar{f}(\alpha^1, \alpha^2, t) = \lim_{\Delta m \to 0} \frac{\bar{F}}{\Delta m} = \frac{d\bar{F}}{dm} = \frac{1}{\rho}\frac{d\bar{F}}{d\sigma}. \tag{5.60}$$

The resultant stress vectors $\bar{R}_i$ acting upon the differential element are found to be

$$\bar{R}_1 = -\left(T^{11}\bar{e}_1 + T^{12}\bar{e}_2\right)\sqrt{\overset{*}{a}_{22}}d\alpha^2, \quad \bar{R}_2 = -\left(T^{21}\bar{e}_1 + T^{22}\bar{e}_2\right)\sqrt{\overset{*}{a}_{11}}d\alpha^1,$$

$$-\left(\bar{R}_1 + \frac{\partial\bar{R}_1}{\partial\alpha^1}d\alpha^1\right) = -\left(T^{11}\bar{e}_1 + T^{12}\bar{e}_2\right)\sqrt{\overset{*}{a}_{22}}d\alpha^2 - \frac{\partial}{\partial\alpha^1}\left(T^{11}\bar{e}_1 + T^{12}\bar{e}_2\right)\sqrt{\overset{*}{a}_{22}}d\alpha^1 d\alpha^2,$$

$$-\left(\bar{R}_2 + \frac{\partial\bar{R}_2}{\partial\alpha^2}d\alpha^2\right) = -\left(T^{21}\bar{e}_1 + T^{22}\bar{e}_2\right)\sqrt{\overset{*}{a}_{11}}d\alpha^1 - \frac{\partial}{\partial\alpha^2}\left(T^{21}\bar{e}_1 + T^{22}\bar{e}_2\right)\sqrt{\overset{*}{a}_{11}}d\alpha^1 d\alpha^2. \tag{5.61}$$

By applying the law of conservation of momentum to (5.59)–(5.61), for the equation of motion of the soft shell, we get

$$\overset{*}{\rho}\frac{d^2\bar{r}(\alpha^1, \alpha^2, t)}{dt^2} = -\frac{\partial\bar{R}_1}{\partial\alpha^1}d\alpha^1 - \frac{\partial\bar{R}_2}{\partial\alpha^2}d\alpha^2 + \bar{p} + \bar{f}\overset{*}{\rho}, \tag{5.62}$$

where $d^2\bar{r}/dt^2$ is acceleration. By substituting $\bar{R}_i$ and $\overset{*}{\rho}$ given by (5.58) and (5.61) into (5.62), we get

$$\rho\sqrt{a}\frac{d^2\bar{r}}{dt^2} = \frac{\partial}{\partial\alpha^1}\left[\left(T^{11}\bar{e}_1 + T^{12}\bar{e}_2\right)\sqrt{\overset{*}{a}_{22}}\right] + \frac{\partial}{\partial\alpha}\left[\left(T^{21}\bar{e}_1 + T^{22}\bar{e}_2\right)\sqrt{\overset{*}{a}_{11}}\right] + \bar{p}\sqrt{\overset{*}{a}} + \bar{f}\rho\sqrt{a}. \tag{5.63}$$

Let $\bar{G}_i$, $\bar{M}_p$, and $\bar{M}_f$ be the resultant moment vectors acting on the element of the shell defined by

$$\bar{G}_1 = \bar{r} \times \bar{R}_1, \qquad\qquad \bar{G}_2 = \bar{r} \times \bar{R}_2$$

$$-\left(\bar{G}_1 + \frac{\partial \bar{G}_1}{\partial \alpha^1}d\alpha^1\right), \qquad -\left(\bar{G}_2 + \frac{\partial \bar{G}_2}{\partial \alpha^2}d\alpha^2\right) \quad (\alpha^i + d\alpha^i = \text{constant}) \quad (5.64)$$

$$\bar{M}_p = (\bar{r} \times \bar{p})\sqrt{\overset{*}{a}}d\alpha^1 d\alpha^2, \quad \bar{M}_f = (\bar{r} \times \bar{f})\rho\sqrt{\overset{*}{a}}d\alpha^1 d\alpha^2.$$

By assuming that the shell is in equilibrium, the sum of the moments vanishes. Hence,

$$-\frac{\partial \bar{G}_1}{\partial \alpha^1}d\alpha^1 - \frac{\partial \bar{G}_2}{\partial \alpha^2}d\alpha^2 + \bar{M}_p + \bar{M}_q = 0. \tag{5.65}$$

By substituting $\bar{G}_i, \bar{M}_p$, and $\bar{M}_f$ in (5.64), we obtain

$$\left[-\left(\bar{r} \times \frac{\partial \bar{R}_1}{\partial \alpha}d\alpha^1\right) - \left(\bar{r} \times \frac{\partial \bar{R}_2}{\partial \alpha^2}d\alpha^2\right) + (\bar{r} \times \bar{p}) + (\bar{r} \times \bar{f})\right]$$

$$- (\bar{r} \times \bar{R}_1)d\alpha^1 - (\bar{r} \times \bar{R}_2)d\alpha^2 = 0.$$

Further, on use of (5.61), we find

$$\left[\bar{e}_1 \times (T^{11}\bar{e}_1 + T^{12}\bar{e}_2) + \bar{e}_2 \times (T^{21}\bar{e}_1 + T^{22}\bar{e}_2)\right]\sqrt{\overset{*}{a_{11}}}\sqrt{\overset{*}{a_{22}}}d\alpha^1 d\alpha^2$$

$$- \bar{r} \times \underline{\left[-\frac{\partial \bar{R}_1}{\partial \alpha^1}d\alpha^1 - \frac{\partial \bar{R}_2}{\partial \alpha^2}d\alpha^2 + \bar{p} + \bar{f}\right]} = 0. \tag{5.66}$$

Since the underlined term equals zero, we have

$$(\bar{e}_1 \times \bar{e}_2)T^{12} + (\bar{e}_2 \times \bar{e}_1)T^{21} = 0. \tag{5.67}$$

It follows immediately from the above $T^{12} = T^{21}$.

Remarks

1. If a soft shell is parameterized along the principal axes, then $T^{11} = T_1$, $T^{22} = T_2$, $T^{12} = 0$, $\overset{*}{a}_{12} = a_{12} = 0$ and the equation of motion (5.62) takes the simplest form

$$\rho\sqrt{a}\frac{d^2\bar{r}}{dt^2} = \frac{\partial}{\partial \alpha^1}\left[T_1\sqrt{\overset{*}{a_{22}}}\bar{e}_1\right] + \frac{\partial}{\partial \alpha^2}\left[T_2\sqrt{\overset{*}{a_{11}}}\bar{e}_2\right] + \bar{p}\sqrt{\overset{*}{a}} + \bar{f}\rho\sqrt{a} \tag{5.68}$$

2. During the dynamic process of deformation, different parts of the soft shell may undergo different stress–strain states. The biaxial stress state occurs when $I_1^{(T)} = T_1 + T_2 > 0$, $I_2^{(T)} = T_1 T_2 > 0$, the uniaxial state develops in areas

where $I_1^{(T)} > 0$, $I_2^{(T)} = 0$,, and the zero stress state takes place anywhere in the shell where $I_1^{(T)} = I_2^{(T)} = 0$. The uniaxially stressed area $(T_2 = 0)$ will develop wrinkles oriented along the action of the positive principal membrane force T_1. The equation of motion for the wrinkled area becomes

$$\rho\sqrt{a}\frac{\mathrm{d}^2\bar{r}}{\mathrm{d}t^2} = \frac{\partial}{\partial\alpha_1}\left[T_1\sqrt{\overset{*}{a_{22}}}\bar{e}_1\right] + \bar{p}\sqrt{\overset{*}{a}} + \bar{f}\rho\sqrt{a}. \tag{5.69}$$

To preserve smoothness and continuity of the surface $\overset{*}{S}$, the uniaxially stressed area is substituted by a "smoothed" surface made out of an array of closely packed reinforced fibers. Such approach allows one to use the equations of motion (5.62) throughout the deformed surface.

The governing system of equations of dynamics of the soft shell includes the equations of motion, constitutive relations, and initial and boundary conditions.

5.6 Nets

A special class of soft shell, where discrete reinforced fibers are the main structural and weight bearing elements, is called nets. Depending on engineering design and practical needs, the fibers may remain discrete or embedded in the connective matrix. Although nets have distinct discrete structure, they are modeled as a solid continuum. Since the nets have very low resistance to shear forces, then $T_{12} = 0$ $(S^r \equiv 0)$ and the resultant formulas obtained in the previous paragraphs are valid in modeling nets.

Consider a net with the cell structure of a parallelogram. Let the sides of the cell be formed by two distinct families of reinforced fibers (Fig. 5.4).

Let the undeformed $(S \equiv \overset{0}{S})$ configuration of the net be parameterized by $\overset{*}{\alpha}_1, \overset{*}{\alpha}_2$ coordinates oriented along the reinforced fibers. For force distribution in the net, we have to

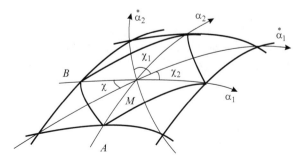

Fig. 5.4 A structural element of the net formed by two distinct types of reinforced fibers. Used with permission from Cambridge University Press

1. Find the stretch ratios $\lambda_1^r = \overset{*}{\lambda}_1$, $\lambda_2^r = \overset{*}{\lambda}_2$, using (5.32)
2. Calculate forces in reinforced fibers: T_1^r, T_2^r,
3. Find the membrane forces in terms of S-configuration

$$
\tilde{T}^{11} = \frac{\lambda_1}{\sin \overset{0}{\chi} \lambda_2} \left(T_1^r \frac{\lambda_2^r}{\lambda_1^r} \sin^2(\overset{0}{\chi} - \overset{0}{\psi}) + T_2^r \frac{\lambda_1^r}{\lambda_2^r} \cos^2(\overset{0}{\chi} - \overset{0}{\psi}) \right),
$$

$$
\tilde{T}^{12} = \frac{1}{\sin \overset{0}{\chi}} \left(T_1^r \frac{\lambda_2^r}{\lambda_1^r} \sin^2(\overset{0}{\chi} - \overset{0}{\psi}) \sin \overset{0}{\psi} - T_2^r \frac{\lambda_1^r}{\lambda_2^r} \cos^2(\overset{0}{\chi} - \overset{0}{\psi}) \cos \overset{0}{\psi} \right), \quad (5.70)
$$

$$
\tilde{T}^{22} = \frac{\lambda_2}{\sin \overset{0}{\chi} \lambda_1} \left(T_1^r \frac{\lambda_2^r}{\lambda_1^r} \sin^2 \overset{0}{\psi} + T_2^r \frac{\lambda_1^r}{\lambda_2^r} \cos^2 \overset{0}{\psi} \right).
$$

The principal membrane forces and their directions are found from (5.54) and (5.55) by putting $T^{12} = S^r = 0$, $T^{11} = T_1^r$, $T^{22} = T_2^r$

$$
T_{1,2} = \frac{(T_1^r + T_2^r) \pm \sqrt{(T_1^r - T_2^r)^2 + 4T_1^r T_2^r \cos^2 \overset{*}{\chi}}}{2 \sin \overset{*}{\chi}},
$$

$$
\tan 2\overset{*}{\psi}_1 = \frac{T_2^r \sin 2\overset{*}{\chi}}{T_1^r + T_2^r \cos 2\overset{*}{\chi}}, \quad (5.71)
$$

$$
\overset{*}{\psi}_2 = \overset{*}{\psi}_1 + \pi/2.
$$

In particular,

1. If $\overset{*}{\chi} = \pi/2$ then $\overset{*}{\psi}_1 = 0$, $T_1 = T_1^r$, $T_2 = T_2^r$,

2. If $T_1^r = 0$ then $\overset{*}{\psi}_1 = 0$, $T_1 = \frac{T_1^r}{\sin \overset{*}{\chi}}$, $T_2 = 0$,

3. If $T_2^r = 0$ then $\overset{*}{\psi}_1 = \overset{*}{\chi}$, $T_1 = \frac{T_2^r}{\sin \overset{*}{\chi}}$, $T_2 = 0$.

5.7 Corollaries of the Fundamental Assumptions

We conclude this chapter with several corollaries of the fundamental assumptions stated at the beginning of Chapter 4 that are specific to thin soft shells.

1. The zero-flexural rigidity state is natural and unique to thin soft shells in contrast to thin elastic shells with finite bending rigidity.
2. Soft shells do not resist compression forces and thus $T_1 \geq 0$, $T_2 \geq 0$ and $I_1^{(T)} \geq 0$, $I_2^{(T)} \geq 0$.
3. Shear membrane forces are significantly smaller compared to stretch forces, $T_{12} \approx 10^{-3} \max T_{ii}$.

4. Areas of the soft shell, where $\Lambda_1 \leq 1$ and $\Lambda_2 \leq 1$, attain multiple configurations and are treated as the zero-stressed areas.
5. Stress states of the soft shell are classified as follows: (1) biaxial, if $T_1 > 0$, $T_2 > 0$, $(I_1^{(T)} > 0, I_2^{(T)} > 0)$, (2) uniaxial, if either $T_1 = 0$, $T_2 > 0$ or $T_1 > 0$, $T_2 = 0$, $(I_1^{(T)} > 0, I_2^{(T)} = 0)$,   and   (3) unstressed, if $T_1 = 0$ and $T_2 = 0$, $(I_1^{(T)} = I_2^{(T)} = 0)$.
6. Constitutive relations for the uniaxial stress–strain state are functions of either Λ_1 or Λ_2 and empirical mechanical constants c_m given by

$$
\begin{aligned}
T_1 &= f_1(\Lambda_1, V^s, c_1, ...c_m, Z_{ij}) \quad \text{for} \quad \Lambda_1 \geq 1, \ \Lambda_2 < 1, \\
T_2 &= f_2(\Lambda_2, V^s, c_1, ...c_m, Z_{ij}) \quad \text{for} \quad \Lambda_1 < 1, \ \Lambda_2 \geq 1.
\end{aligned}
\tag{5.72}
$$

7. Constitutive relations for the in-plane biaxial state, $\Lambda_1 > 1$, $\Lambda_2 > 1(T_1 > 0$, $T_2 > 0)$ have the form

$$
\begin{aligned}
T_1 &= F_1(\Lambda_1, \Lambda_2, V^s, c_1, ...c_m, Z_{ij}), \\
T_2 &= F_2(\Lambda_1, \Lambda_2, V^s, c_1, ...c_m, Z_{ij}), \\
\psi &= \psi(\Lambda_1, \Lambda_2, V^s, c_1, ...c_m, Z_{ij}).
\end{aligned}
\tag{5.73}
$$

In general $f_n(...) \neq F_n(...)$, however, $f_n(...)$ can be defined uniquely if $F_n(...)$ is known.

For example, Corollary 6 of the fundamental assumptions for the nets is given by

$$
\begin{aligned}
I_1^{(T)} &= T_1 + T_2 = \tfrac{1}{\sin \chi} \left(T_1^r + T_2^r\right) \geq 0, \\
I_2^{(T)} &= T_1 T_2 = T_1^r T_2^r \geq 0.
\end{aligned}
$$

Exercises

1. Soft shells (nets) do not have a definite configuration without a priori defined internal/external load. Therefore, the method of calculating of stress–strain distribution in them differs from the ones used in the thin elastic shell theory. What are the ways of calculating stress–strain distribution in soft shells?
2. For the pregnant uterus to be modeled as a soft thin shell, it should satisfy fundamental assumptions as above. What kind of in vivo and in vitro experiments are needed to confirm the assumptions?
3. Verify (5.14) and (5.15).
4. Confirm the relations given by (5.24).
5. Verify (5.29) and (5.30).
6. Verify (5.37).
7. Verify (5.46).
8. Confirm (5.50).

Chapter 6
Continual Model of the Myometrium

It is through science that we prove, but through intuition that we discover.

H. Poincare

6.1 Basic Assumptions

Soft fabrics are different from other engineering materials. They are composite structures whose strength is primarily provided by woven textile and yarn, while the coating guarantees the jointing ability. This results in materials that have low shear stiffness and remarkable tensile properties. Their orthotropic behavior is dictated by micromechanics of weaving: the warp and fill.[1] Different weaves have different mechanical qualities which are mainly governed by the contributions of the yarn and the initial state of crimp in the weave. In general, stretching at service-level stresses is dominated by the weave crimp rather than strain of the yarn fibers. The coating employed in most fabrics tends to attenuate this behavior, especially for transient changes in strain. This leads to a response similar to membranes.

The exact orientation of smooth muscle fibers in the uterus remains controversial. Numerous in vivo and in vitro experimental findings suggest: (1) the two counter rotating spiral muscle fiber arrangement (Goerttler 1968), (2) a circular fiber orientation in the stratum subvasculare only with highly disordered fibers in the rest of the myometrium (Weiss et al. 2006), and (3) a largely interwoven continuous bundle type of organization without definite layers (Ramsey 1994). Studies, involving magnetic resonance imaging, have clearly demonstrated the existence of distinct inner and outer muscular zones within human myometrium. However, the above considerations are based on investigations of nonpregnant uteri and might be inappropriate in the pregnant state.

---

[1]The warp is the direction of the yarn which is spooled out lengthwise in the weaving machine, and the fill or weft is the yarn that runs across.

R.N. Miftahof and H.G. Nam, *Biomechanics of the Gravid Human Uterus*,
DOI 10.1007/978-3-642-21473-8_6, © Springer-Verlag Berlin Heidelberg 2011

Recent three-dimensional structure analysis of the term-pregnant human uterus suggested smooth muscle fasciculi as functional units of the organ. They merge, dichotomize, and intertwine with each other to form an interlacing network (Young and Hession 1999). It is reasonable to assume that the myometrium is a nonlinear heterogenous syncytium and possesses general properties of curvilinear orthotropy with axis of anisotropy defined by the orientation of uterine smooth muscle fasciculi. Being heterogeneous, nonlinear, viscoelastic, incompressible composites, myometrium defies simple material models. Accounting for these particulars in a constitutive model and both experimental evaluations is a great challenge.

The complete theoretical formulation of a mathematical model of the tissue is best achieved with application of the principles of thermodynamics supported by extensive experimentation. The advantage of such approach is that it employs generalized quantities that allow tackling specific problems related to discrete morphological structure of the tissue and the continuum scale of description, which is typically ~1 μm. We shall base our further derivations on the following basic assumptions (Usik 1973):

1. The biomaterial (myometrium) is a two-phase, multicomponent, mechano-chemically active, anisotropic medium; phase 1 comprises the smooth muscle fasciculi, and phase 2 all remaining structures.
2. The phase interfaces are semi-permeable.
3. Active forces of contraction–relaxation produced by smooth muscle are the result of intracellular mechanochemical reactions; the reactions run in a large number of small loci that are evenly distributed throughout the whole volume of the tissue; the sources of chemical reagents are uniformly dispersed within the volume of the biocomposite and are ample.
4. The biocomposite endows properties of general curvilinear anisotropy and viscoelasticity; the viscous properties are due to smooth muscle fiber mechanics and the elastic properties depend mainly on the collagen and elastin fibers.
5. There are no temperature and/or deformation gradients within the tissue.
6. The biocomposite is incompressible and statistically homogeneous.

6.2   Model Formulation

All derivations to follow will be obtained for the averaged parameters. Here we adopt the following notation: the quantities obtained by averaging over the volume of a particular phase are contained in the angle brackets, and those free of brackets are attained by averaging over the entire volume.

Let ρ be the mean density of the tissue. The partial density of the ζth substrate ($\zeta = \overline{1, n}$) in the β phase ($\beta = 1, 2$) is defined as

$$\rho_\zeta^\beta = m_\zeta^\beta / {}^m \breve{V},$$

where m_ζ^β is the mass of the ζth substrate, $^m\breve{V}$ is the total elementary volume of the tissue $^m\breve{V} = \sum_{\beta=1}^2 {^m\breve{V}}^\beta$. The mass and the effective concentrations of substrates are

$$c_\zeta^\beta = \rho_\zeta^\beta/\rho, \qquad \langle c^\beta \rangle = m_\zeta^\beta/^m\breve{V}^\beta \rho^\beta. \tag{6.1}$$

Assuming $\rho = \langle \rho^\beta \rangle = \text{const}$, we have

$$\langle \rho^\beta \rangle = \sum_{\zeta=1}^n \frac{m_\zeta^\beta}{^m\breve{V}^\beta}. \tag{6.2}$$

Setting $\beta = 1$ we find

$$\langle \rho^1 \rangle = \sum_{\zeta=1}^n \frac{m_\zeta^1}{^m\breve{V}^1} = \sum_{\zeta=1}^n \frac{\rho_\zeta^1 \, {^m\breve{V}}}{^m\breve{V}^1} = \sum_{\zeta=1}^n \frac{c_\zeta^1 \rho \, {^m\breve{V}}}{^m\breve{V}^1} = \rho \sum_{\zeta=1}^n \frac{c_\zeta^1}{\eta},$$

where η is the porosity of phase $\beta (\eta = {^m\breve{V}}/{^m\breve{V}}^\beta)$. It is easy to show that

$$\eta = \sum_{\zeta=1}^n c_\zeta^1 \equiv c^1. \tag{6.3}$$

Hence the mass c_ζ^β and the effective $\langle c_\zeta^\beta \rangle$ concentrations are interrelated by $c_\zeta^\beta = \eta \langle c_\zeta^\beta \rangle$. The sum of all concentrations c_ζ^β in the medium equals 1 $\left(\sum_{\zeta=1}^n c_\zeta^\beta = 1 \right)$.

Change in the concentration of constituents in different phases is due to the exchange of the matter among phases, external fluxes, chemical reactions, and diffusion. Since chemical reactions run only in phase 2 and the substrates move at the same velocity, there is no diffusion within phases. Hence, the equations of the conservation of mass of the ζ th substrate in the medium is

$$\rho \frac{dc_\zeta^1}{dt} = Q_\zeta^1, \quad \rho \frac{dc_\zeta^2}{dt} = Q_\zeta^2 + \sum_{j=1}^r v_{\zeta j} J_j. \tag{6.4}$$

Here Q_ζ^β is the velocity of influx of the ζth substrate into the phase α, $v_{\zeta j} J_j$ is the rate of ζth formation in the jth chemical reaction $(j = \overline{1, r})$. The quantity $v_{\zeta j}$ is related to the molecular mass M_ζ of the substrate ζ and is analogous to the stoichiometric coefficient in the jth reaction. $v_{\zeta j}$ takes positive values if the substrate is formed and becomes negative if the substrate disassociates. Since the mass of reacting components is conserved in each chemical reaction, we have

$$\sum_{\zeta=1}^n v_{\zeta j} = 0.$$

Assume that there is a flux Q_ζ^β of the matter into: (1) phase 1 from phase 2 or the "distributed" external sources, and (2) phase 2 from phases 1 only. Hence, we have

$$Q_\zeta^1 = -Q_\zeta + Q_\zeta^e, \quad Q_\zeta^2 = Q_\zeta, \tag{6.5}$$

where Q_ζ^e is the flux of distributed sources, Q_ζ is the exchange flux between phases. Applying the incompressibility condition to (6.5), we have: $\sum_{\zeta=1}^n Q_\zeta^e = 0$. Additionally, let also $\sum_{\zeta=1}^n Q_\zeta = 0$

Assuming that the effective concentration of substrates remains constant throughout and neglecting the convective transport of matter within phases, then with the help of (6.5) from (6.4), we obtain

$$\rho \frac{\partial c_\zeta^1}{\partial t} = -Q_\zeta + Q_\zeta^e, \quad \rho \frac{\partial c_\zeta^2}{\partial t} = Q_\zeta + \sum_{j=1}^r v_{\zeta j} J_j. \tag{6.6}$$

The equation of continuity and the conservation of momentum for the tissue treated as a three-dimensional solid in a fixed Cartesian coordinate system is given by

$$\rho \frac{\partial^2 u_i}{\partial t^2} = \frac{\partial \sigma_{ij}}{\partial x_j} + \rho f_i. \tag{6.7}$$

Here u_i are the components of the displacement vector, f_i is the mass force, and σ_{ij} is the stress tensor for the whole medium.

Let $U^{(\beta)}, s^{(\beta)}, \sigma_{ij}^\alpha$ be the free energy, entropy, and stresses of each phase. The following equalities are an extension of the assumption v

$$s = c^1 \langle s^1 \rangle + (1 - c^1) \langle s^2 \rangle$$
$$U = c^1 \langle U^1 \rangle + (1 - c^1) \langle U^2 \rangle$$
$$\sigma_{ij} = c^1 \langle \sigma^1 \rangle_{ij} + (1 - c^1) \langle \sigma^2 \rangle_{ij}.$$

The Gibbs relation for each phase are defined by

$$c^1 \langle U^1 \rangle = U^1{}_0(c_\zeta^1, T) + \frac{1}{2\rho} E_{ij_{lm}} \varepsilon_{ij} \varepsilon_{lm} \tag{6.8}$$

$$(1 - c^1) \langle U^2 \rangle = U_0^2(c_\zeta^2, T) + \frac{1}{2\rho} Y_{ij_{lm}} \varsigma_{ij}^2 \varsigma_{lm}^2 \tag{6.9}$$

$$d\langle U^1 \rangle = \frac{1}{\rho} \langle \sigma^1 \rangle_{ij} d\varepsilon_{ij}^1 - \langle s^1 \rangle dT + \sum_{\varsigma=1}^n \langle \mu_\varsigma^1 \rangle d\langle c_\varsigma^1 \rangle \tag{6.10}$$

$$d\langle U^2 \rangle = \frac{1}{\rho} \langle \sigma^2 \rangle_{ij} d\varsigma_{ij}^2 - \langle s^2 \rangle dT + \sum_{\varsigma=1}^n \langle \mu_\varsigma^2 \rangle d\langle c_\varsigma^2 \rangle \tag{6.11}$$

$$\langle \mu_\varsigma^\beta \rangle = \partial \langle U^\beta \rangle / \partial \langle c_\varsigma^\beta \rangle, \quad \langle s^\beta \rangle = \partial \langle U^\beta \rangle / \partial T,$$

where μ_ζ^β is the chemical potential of the ζth substrate in the β ($\beta = 1, 2$) phase, $\mu_\zeta^\beta = \partial c^\beta \langle U^\beta \rangle / \partial c_\zeta^\beta$, ς_{ij} is the elastic (phase 1) and Δ_{ij} is the viscous (phase 2) part of deformation ($\varepsilon_{ij}^\beta = \varsigma_{ij}^\beta + \Delta_{ij}^\beta$), E_{ijkl}, Y_{ijkl} are the fourth rank tensors, T is temperature. Making use of the identities

$$\partial c^\beta \langle U^\beta \rangle / \partial c_\zeta^\beta = \left\langle \mu_\zeta^\beta \right\rangle + \langle U^\beta \rangle - \sum_{\varsigma=1}^{n} \left\langle \mu_\zeta^\beta \right\rangle \left\langle c_\zeta^\beta \right\rangle.$$

Equations (6.10) and (6.11) can be written in the form

$$d\left(c^1 \langle U^1 \rangle \right) = \frac{1}{\rho} c^1 \langle \sigma^1 \rangle_{ij} d\varepsilon_{ij}^1 - c^1 \langle s^1 \rangle dT + \sum_{\varsigma=1}^{n} \mu_\zeta^1 dc_\zeta^1$$

$$d\left(c^2 \langle U^2 \rangle \right) = \frac{1}{\rho} c^2 \langle \sigma^2 \rangle_{ij} d\varsigma_{ij}^2 - c^2 \langle s^2 \rangle dT + \sum_{\varsigma=1}^{n} \mu_\zeta^2 dc_\zeta^2.$$

$$\tag{6.12}$$

Assuming that the mass sources are present only in phase 1, the general heat flux and the second law of thermodynamics for the tissue are described by

$$dU = \frac{1}{\rho} \sigma_{ij} d\varepsilon_{ij} - sdT - dq' + \frac{1}{\rho} \sum_{\varsigma=1}^{n} \frac{\partial F}{\partial c_\zeta^1} Q_\zeta^e dt, \tag{6.13}$$

$$Tds = dq^e + dq' + \sum_{\varsigma=1}^{n} TS_\zeta^1 \frac{Q_\zeta^e}{\rho} dt, \tag{6.14}$$

$$S_\zeta^1 = \left(\frac{\partial s}{\partial c_\zeta^1} \right)_{T, c_{\vartheta}^1(\vartheta \neq \zeta), \varsigma_{ij}, \varepsilon_{ij}} = \frac{\partial^2 F}{\partial T \partial c_\zeta^1}.$$

Here $U = \sum_{\beta=1}^{2} c^\beta \langle U^\beta \rangle$, $s = \sum_{\beta=1}^{2} c^\beta \langle s^\beta \rangle$, and S_ζ^1 is the partial entropy of the tissue. On use of (6.12) in (6.13), (6.14), the equation of the balance of entropy of the myometrium takes the form

$$\rho \frac{ds}{dt} - \sum_{\zeta=1}^{n} \frac{\partial s}{\partial c_\zeta^1} Q_\zeta^e = -\text{div} \frac{\overline{q}}{T} + \frac{R}{T}, \tag{6.15}$$

$$R = -\frac{\overline{q}}{T} \text{grad } T + \sigma_{ij}^2 \frac{d\Delta_{ij}^2}{dt} + \sum_{\varsigma=1}^{n} \left(\mu_\zeta^1 - \mu_\zeta^2 \right) Q_\zeta - \sum_{j=1}^{r} J_j \Lambda_j, \tag{6.16}$$

$$\Lambda_j = \sum_{\zeta=1}^{n} v_{\varsigma j} \mu_\zeta^2. \tag{6.17}$$

Here $\boldsymbol{R}$ is the dissipative function, Λ_j is the affinity constant of the jth chemical reaction.

Let the generalized thermodynamic forces and the heat flux, $\bar{q}$, be defined by

$$-\frac{1}{T^2}\mathrm{grad}\theta, \quad \frac{1}{T}\frac{\mathrm{d}\Delta_{ij}^2}{\mathrm{d}t}, \frac{\left(\mu_\varsigma^1 - \mu_\varsigma^2\right)}{T}, \quad -\frac{\Lambda_j}{T},$$
$$\rho\mathrm{d}q^{(e)} = -\mathrm{div}\bar{q}\mathrm{d}t. \tag{6.18}$$

Then, the thermodynamic fluxes $\bar{q}, Q_\varsigma^e, Q_\varsigma, J_j$ and stresses σ_{ij}^2 can be expressed as linear functions of the generalized thermodynamic forces

$$q_i = -N_{ij}\frac{\partial T}{\partial x_j},$$

$$\sigma_{kl}^1 = E_{ijkl}\varepsilon_{ij},$$

$$\sigma_{kl}^2 = B_{ijkl}\frac{\mathrm{d}\Delta_{ij}^2}{\mathrm{d}t} - \sum_{\beta=1}^{r}\widetilde{D}_{\beta kl}\Lambda_\beta + \sum_{\alpha=1}^{n}\widetilde{M}_{\alpha kl}\left(\mu_\alpha^1 - \mu_\alpha^2\right),$$

$$J_\beta = D_{\beta ij}\frac{\mathrm{d}\Delta_{ij}^2}{\mathrm{d}t} - \sum_{\gamma=1}^{r}{}^1 l_{\beta\gamma}\Lambda_\beta + \sum_{\alpha=1}^{n}{}^2 l_{\alpha\beta}\left(\mu_\alpha^1 - \mu_\alpha^2\right),$$

$$Q_\alpha = M_{\alpha ij}\Delta_{ij}^2 - \sum_{\beta=1}^{r}{}^2 l_{\alpha\beta}\Lambda_\beta + \sum_{\beta=1}^{n}{}^3 l_{\alpha\beta}\left(\mu_\beta^1 - \mu_\beta^2\right). \tag{6.19}$$

Here ${}^m l_{\alpha\beta,\beta\gamma}(m = \overline{1,3})$ are scalars, $\boldsymbol{B}_{ijkl}, \overset{(\sim)}{\boldsymbol{D}}_{nij}, \overset{(\sim)}{\boldsymbol{M}}_{\alpha ij}, \boldsymbol{N}_{ij}$ are the parameters of tensorial nature that satisfy the Onsager reciprocal relations

$$\boldsymbol{B}_{ijkl} = \boldsymbol{B}_{klij}, \widetilde{\boldsymbol{D}}_{nij} = -\boldsymbol{D}_{nij}, \boldsymbol{N}_{ji} = \boldsymbol{N}_{ij}, \widetilde{\boldsymbol{M}}_{\alpha ij} = -\boldsymbol{M}_{\alpha ij}, {}^m l_{\alpha\beta,\beta\gamma} = {}^m l_{\beta\alpha,\gamma\beta},$$

and $\sigma_{kl}^\beta(\beta = 1, 2)$ are the components of the total stress tensor, σ_{kl}, of the biocomposite

$$\sigma_{kl} = E_{ijkl}\varepsilon_{ij}^1 + E_{ijkl}^{ve}\varepsilon_{ij}^2,$$

where E_{ijkl}, E_{ijkl}^{ve} are the tensors of elastic and viscous characteristics ($E_{ijkl}^{ve} = E_{klij}^{ve}$) of the myometrium, i.e., collagen and elastin fibers of the stroma and smooth muscle elements, respectively. The term $\sum_{\beta=1}^{r}\widetilde{D}_{\beta kl}\Lambda_\beta$ is the biological input – "biofactor," Z_{kl}, and accounts for biological processes, e.g., electromechanical, chemical, remodeling, aging, etc., in the tissue. With the help of (6.17) and remembering that $\left\langle\mu_\varsigma^\beta\right\rangle = \partial\langle U^\beta\rangle/\partial\left\langle c_\varsigma^\beta\right\rangle$, it is found to be

$$Z_{kl} := -\sum_{\beta=1}^{r}\widetilde{D}_{\beta kl}\Lambda_\beta = \sum_{\beta=1}^{r}\sum_{\alpha=1}^{n}D_{\beta kl}v_{\alpha\beta}\frac{\partial U^2}{\partial c_\varsigma^2}. \tag{6.20}$$

Thus, the system, including equations of conservation of momentum (6.7) with σ_{kl}^{β} defined by (6.19); conservation of mass of the reacting components (6.4) together with (6.16) and Q_α defined by (6.19); heat influx which is obtained on substituting $s = -\partial U / \partial T$ into (6.16) and the explicit expression for $\bar{q}$, describes the continual mechanical model of myometrium. However, it does not guarantee required relationships between in-plane forces and deformations in the soft shell. To establish the missing link, recall the second Kirchhoff–Love hypothesis, which states that the normal stress σ_{33} is significantly smaller compared to $\sigma_{ij}(i,j=1,2)$. Then, the terms containing σ_{33} can be eliminated. In general, the end formulas are very bulky and are not given here. In applications though, depending on a specific tissue, the formulas can be simplified to a certain degree and even take a compact form.

6.3 Biofactor Z_{kl}

The fundamental function of the gravid uterus during labor and delivery is closely related to electromechanical wave processes and the coordinated propagation of the waves of contraction–relaxation in the organ.

Consider smooth muscle syncytia to be electrically excitable biological medium, (Plonsey and Barr 1984). Applying Ohm's law we have

$$\bar{J}_i = -\left(\hat{g}_{i1} \frac{\partial \Psi_i}{\partial x_1} \bar{e}_1 + \hat{g}_{i2} \frac{\partial \Psi_i}{\partial x_2} \bar{e}_2 \right), \tag{6.21}$$

$$\bar{J}_o = -\left(\hat{g}_{o1} \frac{\partial \Psi_o}{\partial x_1} \bar{e}_1 + \hat{g}_{o2} \frac{\partial \Psi_o}{\partial x_2} \bar{e}_2 \right), \tag{6.22}$$

where $\bar{J}_i, \bar{J}_o$ are the intracellular (i) and extracellular (o) currents, Ψ_i, Ψ_o are the scalar electrical potentials, $\hat{g}_{ij}, \hat{g}_{oj}(j=1,2)$ are the conductivities, and $\bar{e}_1, \bar{e}_2$ are the unit vectors in the directions of α_1, α_2 coordinate lines. Both cellular spaces are coupled through the transmembrane current I_{ml} and potential V_m as

$$I_{ml} = -div\bar{J}_i = div\bar{J}_o, \tag{6.23}$$

$$V_m = \Psi_i - \Psi_o. \tag{6.24}$$

Substituting (6.21), (6.22) into (6.23), we get

$$I_{ml} = \hat{g}_{i1} \frac{\partial^2 \Psi_i}{\partial \alpha_1^2} \bar{e}_1 + \hat{g}_{i2} \frac{\partial^2 \Psi_i}{\partial \alpha_2^2} \bar{e}_2, \tag{6.25}$$

$$I_{ml} = -\hat{g}_{o1} \frac{\partial^2 \Psi_o}{\partial \alpha_1^2} \bar{e}_1 + \hat{g}_{o2} \frac{\partial^2 \Psi_o}{\partial \alpha_2^2} \bar{e}_2. \tag{6.26}$$

Equating (6.25) and (6.26), we find

$$(\hat{g}_{i1} + \hat{g}_{o1})\frac{\partial^2 \Psi_i}{\partial \alpha_1^2} + (\hat{g}_{i2} + \hat{g}_{o2})\frac{\partial^2 \Psi_i}{\partial \alpha_2^2} = \hat{g}_{o1}\frac{\partial^2 V_m}{\partial \alpha_1^2} + \hat{g}_{o2}\frac{\partial^2 V_m}{\partial \alpha_2^2}. \tag{6.27}$$

Solving (6.27) for Ψ_i, we obtain

$$\Psi_i = \frac{1}{4\pi}\iint \left(\frac{\hat{g}_{o1}}{\hat{g}_{i1}+\hat{g}_{o1}}\frac{\partial^2 V_m}{\partial X'^2} + \frac{\hat{g}_{o2}}{\hat{g}_{i2}+\hat{g}_{o2}}\frac{\partial^2 V_m}{\partial Y'^2}\right) + \left[\log\left((X-X')^2 + (Y-Y')^2\right)\right]dX'dY',$$

where the following substitutions are used: $X = \alpha_1/\sqrt{\hat{g}_{i1}+\hat{g}_{o1}}$, $Y = \alpha_2/\sqrt{\hat{g}_{i2}+\hat{g}_{o2}}$. Here the integration variables are primed, and the unprimed variables indicate the space point (α_1', α_2') at which Ψ_i is evaluated. The reverse substitutions of X and Y gives

$$\begin{aligned}
\Psi_i = \frac{1}{4\pi}\iint &\left(\hat{g}_{o1}\frac{\partial^2 V_m}{\partial X'^2} + \hat{g}_{o2}\frac{\partial^2 V_m}{\partial Y'^2}\right) \\
&+ \left[\log\left(\frac{(\alpha_1-\alpha_1')^2}{\hat{g}_{i1}+\hat{g}_{o1}} + \frac{(\alpha_2-\alpha_2')^2}{\hat{g}_{i2}+\hat{g}_{o2}}\right)\right]\frac{d\alpha_1' d\alpha_2'}{\sqrt{(\hat{g}_{i1}+\hat{g}_{o1})(\hat{g}_{i2}+\hat{g}_{o2})}}.
\end{aligned} \tag{6.28}$$

Introducing (6.28) into (6.25), after some algebra we obtain

$$\begin{aligned}
I_{m1} = \frac{\tilde{\mu}_1 - \tilde{\mu}_2}{2\pi G(1+\tilde{\mu}_1)(1+\tilde{\mu}_2)}\iint &\left(\hat{g}_{o1}\frac{\partial^2 V_m}{\partial X'^2} + \hat{g}_{o2}\frac{\partial^2 V_m}{\partial Y'^2}\right) \\
\times &\left[\left(\frac{(\alpha_1-\alpha_1')^2}{G_1} - \frac{(\alpha_2-\alpha_2')^2}{G_2}\right) \Big/ \left(\frac{(\alpha_1-\alpha_1')^2}{G_1} + \frac{(\alpha_2-\alpha_2')^2}{G_2}\right)^2\right]d\alpha_1' d\alpha_2',
\end{aligned} \tag{6.29}$$

here

$$G_1 = \hat{g}_{i1} + \hat{g}_{o1}, \quad G_2 = \hat{g}_{i2} + \hat{g}_{o2},$$
$$G = \sqrt{G_1 G_2}, \quad \tilde{\mu}_1 = \hat{g}_{o1}/\hat{g}_{i1}, \quad \tilde{\mu}_2 = \hat{g}_{o2}/\hat{g}_{i2}.$$

Substituting (6.28) into (6.25), we find the contribution of an ε-neighborhood of $(\alpha_1' = 0, \alpha_2' = 0)$ to I_{m1}. Using the transformations given by $X = \alpha_1/\sqrt{\hat{g}_{i1}}$, $Y = \alpha_2/\sqrt{\hat{g}_{i2}}$, we find

$$\begin{aligned}
I_{m2} = \frac{\sqrt{\hat{g}_{i1}\hat{g}_{i2}}}{4\pi G}&\left(\hat{g}_{o1}\frac{\partial^2 V_m}{\partial X'^2} + \hat{g}_{o2}\frac{\partial^2 V_m}{\partial Y'^2}\right)_{\alpha_1'=\alpha_2'=0} \\
&\times \int \nabla^2\left[\log\left(\frac{X'^2}{G_1/\hat{g}_{i1}} + \frac{Y'^2}{G_2/\hat{g}_{i2}}\right)\right]d\alpha_1' d\alpha_2'.
\end{aligned} \tag{6.30}$$

Here ∇^2 is the Laplace operator. Applying the divergence theorem and performing the gradient operation, the integral in (6.30) is converted to a line integral

$$\int \frac{(2X'\hat{g}_{i1}/G_1)\bar{e}_1 + (2Y'\hat{g}_{i2}/G_2)\bar{e}_2}{(X'^2\hat{g}_{i1}/G_1) + (Y'^2\hat{g}_{i2}/G_2)} \cdot \boldsymbol{n}dC', \qquad (6.31)$$

where dC' is an element of the ε-contour. The result of integration yields

$$I_{m2} = \left(\hat{g}_{o1}\frac{\partial^2 V_m}{\partial \alpha'_1 2} + \hat{g}_{o2}\frac{\partial^2 V_m}{\partial \alpha'_2 2}\right)\left(\frac{\hat{g}_{12}}{G_2} + \frac{2(\tilde{\mu}_1 - \tilde{\mu}_2)}{\pi(1+\tilde{\mu}_1)(1+\tilde{\mu}_2)}\tan^{-1}\sqrt{\frac{G_1}{G_2}}\right) \qquad (6.32)$$

To simulate the excitation and propagation pattern in the anisotropic smooth muscle syncytium, we employ the Hodgkin–Huxley formalism described by

$$C_m\frac{\partial V_m}{\partial t} = -(I_{m1} + I_{m2} + I_{ion}),$$

where C_m is the membrane capacitance, I_{ion} is the total ion current through the membrane. Substituting expressions for I_{m1} and I_{m2} given by (6.29), (6.32), we obtain

$$C_m\frac{\partial V_m}{\partial t} = -\frac{\tilde{\mu}_1 - \tilde{\mu}_2}{2\pi G(1+\tilde{\mu}_1)(1+\tilde{\mu}_2)}\iint\left(\hat{g}_{o1}\frac{\partial^2 V_m}{\partial X'^2} + \hat{g}_{o2}\frac{\partial^2 V_m}{\partial Y'^2}\right)$$

$$\times\left[\left(\frac{(\alpha_1-\alpha'_1)^2}{G_1}\frac{(\alpha_2-\alpha'_2)^2}{G_2}\right)\bigg/\left(\frac{(\alpha_1-\alpha'_1)^2}{G_1} + \frac{(\alpha_2-\alpha'_2)^2}{G_2}\right)^2\right]d\alpha'_1 d\alpha'_2$$

$$-\left(\hat{g}_{o1}\frac{\partial^2 V_m}{\partial \alpha'_1 2} + \hat{g}_{o2}\frac{\partial^2 V_m}{\partial \alpha'_2 2}\right)\left(\frac{\hat{g}_{12}}{G_2} + \frac{2(\tilde{\mu}_1-\tilde{\mu}_2)}{\pi(1+\tilde{\mu}_1)(1+\tilde{\mu}_2)}\tan^{-1}\sqrt{\frac{G_1}{G_2}}\right)$$

$$-I_{ion},$$

$$(6.33)$$

where I_{ion} is the function depending on the type and ion channel properties of the biological tissue.

In the case of electrical isotropy, $\tilde{\mu}_1 = \tilde{\mu}_2 = \tilde{\mu}$, the integral in (6.33) vanishes and we get

$$C_m\frac{\partial V_m}{\partial t} = -\frac{1}{(1+\tilde{\mu})}\left(\hat{g}_{o1}\frac{\partial^2 V_m}{\partial \alpha_1^2} + \hat{g}_{o2}\frac{\partial^2 V_m}{\partial \alpha_2^2}\right) - I_{ion}. \qquad (6.34)$$

Finally, the constitutive relations of mechanochemically active electrogenic myometrium include equations of the continual mechanical model as described above, (6.33) and/or (6.34). The system is closed by formulating the free energy, ion currents, initial and boundary conditions, and the function $Z_{ij} = Z_{ij}(V_m, \mu_i, \hat{g}_{ij}, \hat{g}_{oj})$.

It is noteworthy that the closed system of equations describes the development of forces in absence of active strains and vice versa, a condition which is unique to all biological materials.

Models are generally evaluated for the degree of parameters and constants involved as well as for their accurate and meaningful experimental determination. The constitutive model as described above includes $6 + n + r + \frac{1}{2}(n^2 + r^2 + n + r)$ independents scalar quantities. To obtain their experimental estimates will always remain a great challenge. Therefore, simplified models and descriptors are required to serve specific needs of the investigator. While phenomenological constitutive models are able to fit the experimental data with a high degree of accuracy, they are limited in that they do not give the insight into the underlying cause to the particulars of mechanical behavior. Fine molecular and structure-based models help avoid such ambiguities and are able to reveal the intricacies of functions of tissues. However, they are beyond the scope of this book.

6.4 Special Cases

Constitutive relations for soft biological tissues are usually obtained along structurally preferred directions that are defined by the orientation of reinforced smooth muscle, collagen and elastin fibers, and thus, make them easy to use in calculations.

However, if constitutive relations are obtained in the directions different from the actual parameterization of the shell, the task then is to calculate membrane forces in the principal directions. Consider two typical situations:

Case 1. Constitutive relations are given by (5.73). Then,

1. From (5.39) and (5.40), we calculate the principal deformations Λ_1, Λ_2 and the angle φ^*_1,
2. Using (5.73), we compute the principal membrane forces T_1 and T_2 and the angle ψ,
3. Finally, setting $\overset{*}{\chi}_1 = \pi/2, \overset{*}{\chi}_2 = \psi, \overset{*}{T}^{11} = T_1, \overset{*}{T}^{22} = T_2, \overset{*}{T}^{12} = 0$ in (5.49), we find

$$
\begin{aligned}
\tilde{T}^{11} &= \{T_1\sin^2(\chi - \psi) + T_2\cos^2(\chi - \psi)\}/\sin\chi, \\
\tilde{T}^{12} &= \{T_1 \sin\psi \sin(\chi - \psi) - T_2 \cos\psi \cos(\chi - \psi)\}/\sin\chi, \\
\tilde{T}^{22} &= \{T_1\sin^2\psi + T_2\cos^2\psi\}/\sin\chi.
\end{aligned}
\qquad (6.35)
$$

Case 2. Constitutive relations are formulated for the orientation of reinforced fibers, superscript (r),

$$
\begin{aligned}
T_1^r &= F_1^r(\lambda_1^r, \lambda_2^r, \gamma^r, c_1, ...c_m, Z_{ij}), \\
T_2^r &= F_2^r(\lambda_1^r, \lambda_2^r, \gamma^r, c_1, ...c_m, Z_{ij}), \\
S^r &= S^r(\lambda_1^r, \lambda_2^r, \gamma^r, c_1, ...c_m, Z_{ij}).
\end{aligned}
\qquad (6.36)
$$

Let $\overset{*}{\alpha}_i \in \overset{*}{S}$ be an auxillary orthogonal coordinate system oriented with respect to a set of reinforced fibers by $\overset{*}{\psi}$. Then,

1. Setting $\chi_1 = \pi/2, \chi_2 = \psi, \overset{*}{\chi}_1 = \pi/2 - \gamma^r$ in (5.32), where $\lambda_1^r := \overset{*}{\lambda}_1, \lambda_2^r := \overset{*}{\lambda}_2$, for the stretch ratios and the shear angle γ^r, we have

$$\lambda_1^r = \left(\lambda_1^2 \sin^2(\overset{0}{\chi} - \overset{0}{\psi}) + \lambda_2^2 \sin^2 \overset{0}{\psi} + 2\lambda_1\lambda_2 \cos \overset{0}{\chi} \sin(\overset{0}{\chi} - \overset{0}{\psi}) \sin \overset{0}{\psi} \right)^{1/2} \Big/ \sin \overset{0}{\chi},.$$

$$\gamma^r = \sin^{-1}\left(\frac{1}{\lambda_1\lambda_2 \sin^2 \overset{0}{\chi}} - \left(\frac{1}{2}\lambda_1^2 \sin 2(\overset{0}{\chi} - \overset{0}{\psi}) + \frac{1}{2}\lambda_2^2 \sin 2\overset{0}{\psi} + \lambda_1\lambda_2 \cos \overset{0}{\chi} \sin(\overset{0}{\chi} - \overset{0}{\psi}) \right) \right)$$

$$\lambda_2^r = \left(\lambda_1^2 \cos^2(\overset{0}{\chi} - \overset{0}{\psi}) + \lambda_2^2 \cos^2 \overset{0}{\psi} - 2\lambda_1\lambda_2 \cos(\overset{0}{\chi} - \overset{0}{\psi}) \cos \overset{0}{\chi} \cos \overset{0}{\psi} \right)^{1/2} \Big/ \sin \overset{0}{\chi}.$$

$$(6.37)$$

2. Using (5.73) we find T_1^r, T_2^r and S^r,
3. The angle $\overset{*}{\psi}$ is found from (6.32) by putting $\overset{*}{\chi}_1 = \psi, \chi_2 = 0, \overset{}{\underset{1}{\lambda}} = \lambda_1, \overset{}{\underset{2}{\lambda}} = \lambda_1^r$,

$$\overset{*}{\psi} = \cos^{-1}\left(\frac{1}{\lambda_1^r \sin \overset{0}{\chi}} \left(\lambda_1 \sin(\overset{0}{\chi} - \overset{0}{\psi}) + \lambda_2 \cos \overset{0}{\chi} \sin \overset{0}{\psi} \right) \right), \qquad (6.38)$$

4. Setting $\overset{*}{\chi}_1 = \pi/2 - \gamma^r, \chi_2 = \overset{*}{\psi}, T_1^r := \tilde{T}^{11}, T_2^r := \tilde{T}^{22}, S^r := \tilde{T}^{12}$ in (5.49), we obtain

$$\tilde{T}^{11} = \left\{ T_1^r \sin^2 (\overset{*}{\chi} - \overset{*}{\psi}) + T_2^r \cos^2 (\overset{*}{\chi} - \overset{*}{\psi} + \gamma^r) \right.$$

$$\left. - 2S^r \cos(\overset{*}{\chi} - \overset{*}{\psi} + \gamma^r) \sin(\overset{*}{\chi} - \overset{*}{\psi}) \right\} \Big/ \sin \overset{*}{\chi} \cos \gamma^r,$$

$$\tilde{T}^{12} = \left\{ T_1^r \sin \mu \sin(\overset{*}{\chi} - \overset{*}{\psi}) - T_2^r \cos(\overset{*}{\psi} + \gamma^r) \cos(\overset{*}{\chi} - \overset{*}{\psi} + \gamma^r) \right.$$

$$\left. + S^r \left[\sin(\overset{*}{\chi} - 2\overset{*}{\psi} + \gamma^r) - \cos \overset{*}{\chi} \sin \gamma^r \right] \right\} \Big/ \sin \overset{*}{\chi} \cos \gamma^r,$$

$$\tilde{T}^{22} = \left\{ T_1^r \sin^2 \overset{*}{\psi} - 2S^r \sin(\overset{*}{\psi} - \gamma^r) \sin \overset{*}{\psi} + T_2^r \cos^2 (\overset{*}{\psi} - \gamma^r) \right\} \Big/ \sin \overset{*}{\chi} \cos \gamma^r. \quad (6.39)$$

Formulas (6.39) can be written in more concise form if we introduce generalized forces defined by $N^{ik} = T^{ik} \dfrac{\lambda_k}{\lambda_i}$,

$$\overset{*}{N}^{11} = T_1^r \frac{\lambda_2^r}{\lambda_1^r}, \qquad \overset{*}{N}^{22} = T_2^r \frac{\lambda_1^r}{\lambda_2^r}, \qquad \overset{*}{N}^{12} = S^r.$$

Then, (5.50) takes the form

$$N^{ik} = \frac{1}{\overset{*}{\hat{C}}} \overset{*}{N}{}^{jn} \overset{\wedge}{\hat{C}}{}^{*i}_{j} \overset{\wedge}{\hat{C}}{}^{*k}_{n} . \tag{6.40}$$

Substituting $\overset{\wedge}{\hat{C}}{}^{*i}_{j}$ given by (5.16) and (5.18), we find

$$N^{11} = \left\{ \overset{*}{N}{}^{11} \sin^2 (\overset{0}{\chi} - \overset{0}{\chi}_2) + \overset{*}{N}{}^{22} \sin^2(\overset{0}{\chi}_1 + \overset{0}{\chi}_2 - \overset{0}{\chi}) \right.$$

$$\left. -2\overset{*}{N}{}^{12} \sin(\overset{0}{\chi}_1 + \overset{0}{\chi}_2 - \overset{0}{\chi}) \sin(\overset{0}{\chi} - \overset{0}{\chi}_2) \right\} / \sin\overset{0}{\chi} \sin\overset{0}{\chi}_1$$

$$N^{12} = \left\{ \overset{*}{N}{}^{11} \sin\overset{0}{\chi}_2 \sin(\overset{0}{\chi} - \overset{0}{\chi}_2) + \overset{*}{N}{}^{22} \sin(\overset{0}{\chi}_1 + \overset{0}{\chi}_2 - \overset{0}{\chi}) \sin(\overset{0}{\chi}_1 + \overset{0}{\chi}_2) \right. \tag{6.41}$$

$$\left. + \overset{*}{N}{}^{12} \left[\cos(\overset{0}{\chi}_1 + 2\overset{0}{\chi}_2 - \overset{0}{\chi}) - \cos\overset{0}{\chi} \cos\overset{0}{\chi}_1 \right] \right\} / \sin\overset{0}{\chi} \sin\overset{0}{\chi}_1,.$$

$$N^{22} = \left\{ \overset{*}{N}{}^{11} \sin^2\overset{0}{\chi}_2 - 2\overset{*}{N}{}^{12} \sin(\overset{0}{\chi}_1 + \overset{0}{\chi}_2) \sin\overset{0}{\chi}_2 + \overset{*}{N}{}^{22} \sin^2(\overset{0}{\chi}_1 + \overset{0}{\chi}_2) \right\}$$

Putting $\overset{0}{\chi}_1 = \pi/2$, $\overset{0}{\chi}_2 = \overset{0}{\psi}$ in (6.41), for the membrane forces in terms of the undeformed surface $S(\overset{0}{S} = S)$, we have

$$\tilde{T}^{11} = \frac{\lambda_1}{\lambda_2} \left(T^r_1 \frac{\lambda^r_2}{\lambda^r_1} \sin^2 (\overset{0}{\chi} - \overset{0}{\psi}) + T^r_2 \frac{\lambda^r_1}{\lambda^r_2} \cos^2(\overset{0}{\chi} - \overset{0}{\psi}) - 2S^r \sin 2(\overset{0}{\chi} - \overset{0}{\psi}) \right) / \sin\overset{0}{\chi},$$

$$\tilde{T}^{12} = \left(T^r_1 \frac{\lambda^r_2}{\lambda^r_1} \sin^2 (\overset{0}{\chi} - \overset{0}{\psi}) \sin\overset{0}{\psi} - T^r_2 \frac{\lambda^r_1}{\lambda^r_2} \cos^2 (\overset{0}{\chi} - \overset{0}{\psi}) \cos\overset{0}{\psi} + S^r \sin(\overset{0}{\chi} - 2\overset{0}{\psi}) \right) / \sin\overset{0}{\chi},$$

$$\tilde{T}^{22} = \frac{\lambda_2}{\lambda_1} \left(T^r_1 \frac{\lambda^r_2}{\lambda^r_1} \sin^2 \overset{0}{\psi} + T^r_2 \frac{\lambda^r_1}{\lambda^r_2} \cos^2 \overset{0}{\psi} + S^r \sin 2\overset{0}{\psi} \right) / \sin\overset{0}{\chi} ..$$

$$\tag{6.42}$$

Formulas (6.42) depend only on parameterization of S and the axes of anisotropy and are less computationally demanding compared to (6.38) and (6.39).

Exercises

1. Discuss the validity of basic assumptions (1)–(6).
2. Verify (6.10) and (6.11).
3. Verify (6.29).
4. Confirm (6.32).
5. Verify (6.35) and (6.39).
6. Verify (6.41).

Chapter 7
Models of Synaptic Transmission and Regulation

Divide each difficulty into as many parts as is feasible and necessary to resolve it.

R. Descartes

7.1 System Compartmentalization

Neurohormonal modulation and electrochemomechanical coupling in myometrium, as described in Chap. 1, involve a cascade of chemical processes including synthesis, storage, stimulation, release, diffusion, and binding of various substrates to specific receptors with activation of intracellular second messenger systems and the generation of a variety of physiological responses. Qualitative analysis and quantitative evaluation of the each and every step experimentally is very difficult, and sometimes practically impossible. Therefore, different classes of mathematical models of neurohormonal modulation and synaptic neurotransmission, ranging from the most comprehensive "integrated" – microphysiological – to "reductionist" – deterministic – have been proposed to study intricacies of the processes of neuroendocrine regulations. With microphysiological approach, attempts to reproduce reality in great detail lead to mathematically challenging and computationally demanding tasks. In contrast, deterministic models aim to capture accurately phenomenological behavior of the system. They not only provide macroscopic explanation of complex biophysical processes but are general enough to offer a coherent description of essential biochemical reactions within the unified framework. These models are inherently flexible and can accommodate spatiotemporal and structural interactions into a tractable representation.

Let a myometrial synapse be an open three compartmental system. Compartment 1 comprises presynaptic elements where synthesis and storage of a neurotransmitter or a hormone occurs. In case of cholinergic or adrenergic synapses it corresponds morphologically to a nerve terminal of the unmyelinated axon; for oxytocin, it is comprised of large dense-core vesicles that are found in the corpus luteum and

R.N. Miftahof and H.G. Nam, *Biomechanics of the Gravid Human Uterus*,
DOI 10.1007/978-3-642-21473-8_7, © Springer-Verlag Berlin Heidelberg 2011

placental cells, and prostaglandins $F_{2\alpha}$ and E_2 are synthesized in the decidua, the chorion leave and the amnion.

Ligands, i.e. acetylcholine, adrenalin and oxytocin, are released upon neural stimulation by exocytosis to the synaptic cleft and bloodstream. The common sequence of events involved in the dynamics of their transduction includes:

1. Depolarization of the nerve terminal or cell membrane
2. Influx of extracellular calcium through voltage-gated Ca^{2+} channels
3. Binding of free cytosolic Ca_i^{2+} to transmitter-containing vesicles
4. Release of vesicular/granular stored ligand, L_v, into the synaptic cleft.

Propagation of the wave of depolarization, V^f, in the terminal is accurately described by the modified Hodgkin–Huxley system of equations (Miftahof et al. 2009)

$$C_m^f \frac{\partial V^f}{\partial t} = \frac{1}{2R_a^f} \frac{\partial}{\partial \alpha}\left(d_f^2(\alpha)\frac{\partial V^f}{\partial \alpha}\right) - \left(I_{Na}^f + I_K^f + I_{Cl}^f\right),$$

$$d_f(\alpha) = \begin{cases} d_f, & L^s < \alpha \le L^s - L^s_0 \\ 2d_f & \alpha > L^s - L^s_0, t > 0, \alpha \in (0, L^s). \end{cases} \tag{7.1}$$

Here C_m^f is the specific capacitance of the nerve fiber, R_a^f is membrane resistance, d_f is the cross-sectional diameter of the terminal, α is the Lagrange coordinate, and L^s, L^s_0 are the lengths of the axon and the terminal, respectively. The above ion currents are defined by

$$I_{Na}^f = g_{Na}^f m_f^3 h_f (V^f - V_{Na}^f)$$

$$I_K^f = g_K^f n_f^4 h_f (V^f - V_K^f)$$

$$I_{Cl}^f = g_{Cl}^f (V^f - V_{Cl}^f), \tag{7.2}$$

where $g_{Na}^f, g_K^f, g_{Cl}^f$ are the maximal conductances for Na^+, K^+ and Cl^- currents, respectively, $V_{Na}^f, V_K^f, V_{Cl}^f$ are the equilibrium potentials for the respective ion currents, and m_f, n_f, h_f are the state variables that are calculated from

$$dy/dt = \alpha_y(1-y) - \beta_y y \quad y = \left(m_f, n_f, h_f\right). \tag{7.3}$$

The activation, α_y, and deactivation, β_y, parameters satisfy the following empirical relations

$$\alpha_{m,f} = \frac{0.1T(2.5 - V^f)}{\exp(2.5 - 0.1V^f)}, \qquad \beta_{m,f} = 4T\exp(-V^f/18)$$

$$\alpha_{h,f} = 0.07T\exp(-0.05V^f), \quad \beta_{h,f} = T/(1 + \exp(3 - 0.1V^f))$$

$$\alpha_{n,f} = \frac{0.1T(10 - V^f)}{\exp(1 - 0.1V^f) - 1}, \quad \beta_{n,f} = 0.125T\exp(-0.125V^f). \tag{7.4}$$

Here T is temperature.

The cytosolic calcium turnover is given by

$$\frac{d[Ca_i^{2+}]}{dt} = g_{syn}^{Ca} V^f(t)[Ca_0^{2+}] - k_b[Ca_i^{2+}], \tag{7.5}$$

where g_{syn}^{Ca} is the conductivity of the voltage-gated Ca^{2+} channel at the synaptic end, k_b is the intracellular buffer system constant.

Finally, the release of a stored fraction of the ligand, L_v, is described by the state diagram

$$\frac{d[X_2]}{dt} = -k_0[X_1][X_2], \tag{7.6}$$

Here $X_1 := [Ca_i^{2+}], X_2 := [L_v]$ are concentrations of the cytosolic calcium and vesicular-stored ligand, respectively, k_0 is the rate constant of association of Ca_i^{2+} with calcium-dependent centers on the vesicles, k_1 is the diffusion constant.

At $t = 0$ the nerve axon and the synapse are assumed to be at the resting state and the concentrations of reacting components are known

$$m_f(0) = m_{f,0}, h_f(0) = h_{f,0}, n_f(0) = n_{f,0},$$
$$V^f(0, \alpha) = 0, L_v(0) = L_{v,0}. \tag{7.7}$$

The synapse is excited at the free end ($\alpha = 0$) by the electric impulse of an amplitude $\overset{0}{V^f}$ and a duration t^d, and the presynaptic terminal end ($\alpha = L$) remains unexcited throughout.

$$V^f(0, t) = \begin{cases} \overset{0}{V^f}, & 0 < t < t^d, \\ 0, & t \geq t^d \end{cases} \quad V^f(L, t) = 0. \tag{7.8}$$

The release of prostaglandins $F_{2\alpha}$ and E_2 is mediated by a specific transporter – the multidrug resistance-associated protein 4, a member of the ATP-binding cassette transporter superfamily. However, whether it is the only carrier involved is still unclear. As a first approximation, let the PGs dynamics satisfy

$$\frac{d[PG_i]}{dt} = \frac{B_{tr}([PG_i] - [PG_o])}{K_{tr} + [PG_i]}, PG_i(0) = PG_{io} \tag{7.9}$$

Here B_{tr}, K_{tr} are the parameters that are referred to properties of the transporter, $[PG]_{i,o}$ are the concentrations of prostaglandin inside (subscript i) and outside (o) the cell.

Progesterone is constantly produced and secreted to the bloodstream as an active hormone. Its concentration is tightly regulated and maintained at a constant level. Therefore, we assume that [PR] = constant throughout and is defined a priori.

Upon their release, neurotransmitters and hormones passively diffuse through the synaptic cleft (compartment 2) toward the postsynaptic membrane. The synaptic cleft across many synaptic types has the width of $\approx$12–20 nm. It contains N-cadherin, adhesion molecules, microtubules organized in periodic transsynaptic complexes and roughly regular patches. Their function is only beginning to be understood (Yamagata et al. 2003).

Compartment 3 – the postsynaptic membrane – contains a complex network of surface membrane proteins which modulate the transmission. It involves the following processes:

5. Binding of free ligand in the cleft, L_c, to the G-protein-coupled receptor, R, and its conformational change, $L_c \cdot R$.
6. Active configuration of the receptor, R^* (the $L_c \cdot R^*$ reactive complex is able to produce a biological effect, i.e. the postsynaptic potential, V_{syn}).
7. Binding of $L_c \cdot R^*$ to G protein, the formation of the $L_c \cdot R^* \cdot G$ - complex and the initiation of guanosine diphosphate/guanosine triphosphate (GDP/GTP) exchange.
8. The dissociation of G protein α and $\beta\gamma$ subunits with the subsequent release of G_{act} protein that interacts with downstream effector pathways.
9. Enzymatic, E, clearance of the excess of L_c in the synaptic cleft through the formation of intermediate complexes, $L_c \cdot E$ and ES, and the final metabolite, S.

They are described by the following state diagram (Fig. 7.2).

Here $k_{\pm i}$ are the forward ($+i$) and backward ($-i$) rate constants of chemical reactions, $C_j(j = \overline{1,4})$ are the intermediate complexes, and the meaning of other parameters as described above.

Assuming that:

1. The distribution of reactive substrates is uniform throughout and no chemical gradients exist.
2. The total enzyme concentration does not change over time, E_0 = constant.
3. The total ligand concentration is much larger that the total enzyme concentration.
4. No product is present at the beginning of the reaction, and
5. The maximum rate of chemical reaction occurs when the enzyme is saturated, i.e. all enzyme molecules are tied up with a substrate.

then all reactions are of the first order and satisfy the Michaelis–Menten kinetics. Hence the system of equations for the ligand conversion is given by

$$d\mathbf{X}/dt = \mathbf{A}\mathbf{X}(t), \tag{7.10}$$

where the vector $\mathbf{X}(t) = \left(X_j\right)^{\mathrm{T}} (j = \overline{1,20})$ has the components

$$
\begin{aligned}
&X_1 = [\mathrm{Ca}_i^{2+}], &&X_2 = [L_v], &&X_3 = [L_v \cdot \mathrm{Ca}_i^{2+}], &&X_4 = [L_c], \\
&X_5 = [R], &&X_6 = [R^*], &&X_7 = [L_c \cdot R^*], &&X_8 = [L_c \cdot R], \\
&X_9 = [L_c \cdot R^* \cdot G], &&X_{10} = [R^* \cdot G], &&X_{11} = [A_c \cdot E], &&X_{12} = [G], \\
&X_{13} = [C_1], &&X_{14} = [C_2], &&X_{15} = [C_3], &&X_{16} = [C_4], \\
&X_{17} = [G'], &&X_{18} = [G_{\mathrm{act}}], &&X_{19} = [ES], &&X_{20} = [S].
\end{aligned}
$$

The matrix $\mathbf{A}(a_{ij})(i,j = \overline{1,20})$ has the nonzero elements

$$
\begin{aligned}
&a_{11} = g_{\mathrm{syn}}^{\mathrm{Ca}} V^{\mathrm{f}}(t)[\mathrm{Ca}_0^{2+}], &&a_{12} = -k_0[X_1], &&a_{13} = k_{-0} \\
&a_{22} = -k_0[X_1], &&a_{23} = k_{-0} \\
&a_{32} = k_0[X_1], &&a_{33} = -(k_{-0} + k_1), \\
&a_{43} = k_1, &&a_{44} = k_2([E_0] - [X_8]) + k_6[X_5] + k_{-8}[X_6] + k_{-11}[X_{10}], \\
&a_{47} = k_8, &&a_{48} = k_{-6}, &&a_{49} = k_{11}, \\
&a_{4,11} = k_{-2}, \\
&a_{55} = -(k_{-5} + k_6)[X_4], &&a_{56} = k_5[X_{12}], &&a_{58} = k_{-6}, \\
&a_{65} = k_{-5}, &&a_{66} = -k_{-8}[X_4] - (k_5 + k_{-10}k_{19}[X_{18}])[X_{12}], \\
&a_{67} = k_8[X_{12}], &&a_{6,10} = k_{10}, &&a_{6,16} = k_{17}, \\
&a_{76} = k_{-8}[X_4], &&a_{77} = -(k_{-7} + k_8 + k_9)k_{19}[X_{12}][X_{18}], \\
&a_{78} = k_7, &&a_{79} = k_{-9}k_{12}[X_{40}], &&a_{7,14} = k_{16}, \\
&a_{85} = k_6[X_4], &&a_{87} = k_{-7}[X_{12}], &&a_{88} = -k_{-6} - k_7, \\
&a_{97} = k_9k_{19}[X_{12}][X_{18}], &&a_{99} = -k_{-9} - k_{11} - k_{12}[X_{40}], &&a_{9,10} = k_{-11}[X_4], \\
&a_{10,6} = k_{-10}[X_{12}], &&a_{10,9} = k_{11}, &&a_{10,10} = -(k_{10} + k_{-11})[X_4] - k_{13}[X_{40}], \\
&a_{11,4} = k_2([X_{41}] - [X_8]), &&a_{11,11} = -(k_{-2} + k_3), &&a_{11,19} = k_3, \\
&a_{12,6} = k_{-5} + k_{-8}[X_4], &&a_{12,8} = k_7 &&a_{12,9} = k_{-9}, &&a_{12,10} = k_{10}, \\
&a_{12,12} = -(k_5 + k_{-10}k_{19}[X_{18}])[X_6] - (k_{-7} + k_8 + k_9k_{19}[X_{18}])[X_7], \\
&a_{13,9} = k_{12}[X_{40}], &&a_{13,13} = -k_{14}[X_{39}], &&a_{13,14} = k_{-14}, \\
&a_{14,13} = k_{14}[X_{39}], &&a_{14,14} = -(k_{-14} + k_{16}), \\
&a_{15,10} = k_{13}[X_{40}], &&a_{15,15} = -k_{15}[X_{39}], &&a_{15,16} = k_{-15}, \\
&a_{16,15} = k_{15}[X_{39}], &&a_{16,16} = -(k_{17} + k_{-15}), \\
&a_{17,14} = k_{16}, &&a_{17,16} = k_{17}, &&a_{17,17} = -k_{18}, \\
&a_{18,17} = k_{18}, &&a_{18,18} = -k_{19}, \\
&a_{19,11} = k_3, &&a_{19,19} = -(k_{-3} + k_4), &&a_{19,20} = k_{-4}([X_{41}] - [X_8]), \\
&a_{20,19} = k_4, &&a_{20,20} = -k_{-4}([X_{41}] - [X_8]).
\end{aligned}
$$

Here $[X_{39}]$, $[X_{40}]$, $[X_{41}]$ are given concentrations of *GTP*, *GDP* and E enzymes, respectively. Note that in case of PGF$_{2\alpha}$ and PGE$_2$ their release is described by (7.9).

The generation of the excitatory postsynaptic potential, V_{syn}, is given by (Miftahof et al. 2009)

$$C_p \frac{dV_{syn}}{dt} + V_{syn}(-\Omega[X_9] + R_v^{-1}) = \frac{V_{syn,0}}{R_v}, \qquad (7.11)$$

where C_p is the capacitance of the postsynaptic membrane, R_v is the resistance of the synaptic structures, Ω is the empirical constant, $V_{syn,0}$ is the resting postsynaptic potential.

Given the concentrations of reacting components and the state of the synapse

$$\mathbf{X}(0) = \mathbf{X}_0, \quad V_{syn} = 0, \qquad (7.12)$$

Equations (7.1)–(7.12) provide mathematical formulation of the dynamics of the common pathway of neurotransmission by acetylcholine, adrenalin, oxytocin, and prostaglandins at the myometrial site.

The quantitative assessment of the velocities of reactions shows that the rates, k_{12}, k_{13}, of exchange of G protein for *GDP* at $L_c \cdot R^* \cdot G$ and $R^* \cdot G$ sites, respectively, are significantly smaller compared to the rates of other reactions. Hence, the system (Figs. 7.1 and 7.2) can be viewed as a combination of rapid equilibrium segments interconnected through slow, rate-limiting steps. The characteristic feature of such system is that it attains the steady state when the rapid segment has already reached quasi-equilibrium. This fact allows us to simplify the system (7.10) as follows:

Let $\tilde{f}_1$ and $\tilde{f}_2$ be the fractional concentration factors of the rapid segment "product"-substrates

$$\tilde{f}_1 = [X_9]/[Y], \quad \tilde{f}_2 = [X_{10}]/[Y], \qquad (7.13)$$

where Y is

$$[Y] = [X_9] + [X_{10}] + [X_{12}].$$

Define the association constant as

$$K_1 = \frac{[X_6][X_{12}]}{[X_{10}]}, \quad K_2 = \frac{[X_4][X_{10}]}{[X_9]}, \quad K_3 = \frac{[X_7][X_{10}]}{[X_9]}, \quad K_4 = \frac{[X_4][X_6]}{[X_9]},$$

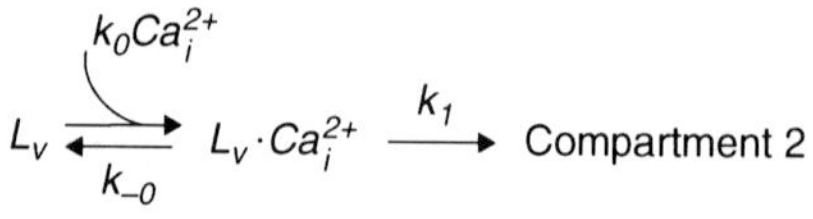

Fig. 7.1 The state diagram of a transmitter release from the presynaptic terminal

$$L_v \underset{k_{-0}}{\overset{k_0 Ca_i^{2+}}{\rightleftharpoons}} L_v \cdot Ca_i^{2+} \overset{k_1}{\longrightarrow} \text{Compartment 2}$$

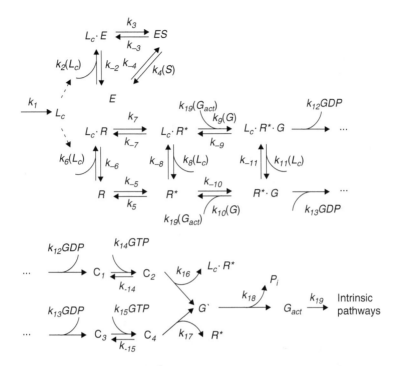

Fig. 7.2 The state diagram of a transmitter conversion at the postsynaptic membrane

after simple algebra, for $\widetilde{f}_1$, we have

$$
\widetilde{f}_1 = \frac{\frac{[X_4][X_6]}{K_1 K_2}}{1 + \frac{[X_4][X_6]}{K_1 K_2} + \frac{[X_6]}{K_1}} = \frac{[X_4][X_6]}{K_1 K_2 + [X_6]([X_4] + K_2)},
$$

$$
\widetilde{f}_2 = \frac{\frac{[X_6]}{K_1}}{1 + \frac{[X_4][X_6]}{K_1 K_2} + \frac{[X_6]}{K_1}} = \frac{K_2[X_6]}{K_{10} K_{11} + [X_6]([X_4] + K_2)}.
$$

(7.14)

Then, concentration distribution equations for the main reactants can be obtained in the form (for details see King and Altman 1956; Cha 1968)

$$
(D)\frac{[Y]}{[Y']} = k_{14}k_{15}k_{16}k_{17}k_{18}k_{19}[X_{39}]^2
$$

$$
(D)\frac{[X_{13}]}{[Y']} = k_{12}k_{15}k_{17}k_{18}k_{19}\widetilde{f}_1[X_{39}](k_{-14} + k_{16})
$$

$$
(D)\frac{[X_{14}]}{[Y']} = k_{12}k_{14}k_{15}k_{17}k_{18}k_{19}\widetilde{f}_1[X_{39}]^2
$$

$$
(D)\frac{[X_{15}]}{[Y']} = k_{11}k_{14}k_{16}k_{18}k_{19}\widetilde{f}_2[X_{39}](k_{-15} + k_{17})
$$

(7.15)

$$
(D)\frac{[X_{16}]}{[Y]} = k_{11}k_{14}k_{15}k_{16}k_{18}k_{19}\widetilde{f}_2[X_{39}]^2
$$

$$
(D)\frac{[X_{17}]}{[Y']} = k_{14}k_{15}k_{16}k_{17}\widetilde{f}_2[X_{39}]^2(k_{12}\widetilde{f}_1 + k_{11}\widetilde{f}_2)
$$

where

$$[Y'] = [Y] + [X_{13}] + [X_{14}] + [X_{15}] + [X_{16}] + [X_{17}],$$

and D is the sum of all the values on the right side of (7.14). Note that (7.15) are algebraic equations. The initial velocity equation for Y complexes formation is given by

$$d[Y]/dt = k_{18}k_{19}[X_{17}]. \tag{7.16}$$

Finally, substituting X_{17} from (7.15) and making use of (7.14) we get

$$\frac{1}{[Y']}\frac{d[Y]}{dt} = \frac{k_{14}k_{15}k_{16}k_{17}k_{18}k_{19}[X_{39}]^2 \left[\dfrac{k_{12}\frac{[X_4][X_6]}{K_1K_2} + k_{11}\frac{[X_6]}{K_1}}{1 + \frac{[X_4][X_6]}{K_1K_2} + \frac{[X_6]}{K_1}}\right]}{\left\{\begin{array}{l} k_{14}k_{15}k_{16}k_{17}k_{18}k_{19}[X_{39}]^2 + \\[2mm] k_{12}k_{15}k_{17}k_{18}k_{19}[X_{39}](k_{-14} + k_{16})\left[\dfrac{\frac{[X_4][X_6]}{K_1K_2}}{1 + \frac{[X_4][X_6]}{K_1K_2} + \frac{[X_6]}{K_1}}\right] + \\[4mm] + k_{12}k_{14}k_{15}k_{17}k_{18}k_{19}[X_{39}]^2 \left[\dfrac{\frac{[X_4][X_6]}{K_1K_2}}{1 + \frac{[X_4][X_6]}{K_1K_2} + \frac{[X_6]}{K_1}}\right] + \\[4mm] + k_{14}k_{15}k_{16}k_{17}[X_{39}]^2 \left[\dfrac{k_{12}\frac{[X_4][X_6]}{K_1K_2} + k_{11}\frac{[X_6]}{K_1}}{1 + \frac{[X_4][X_6]}{K_1K_2} + \frac{[X_6]}{K_1}}\right] + \\[4mm] + k_{11}k_{14}k_{15}k_{16}k_{18}k_{19}[X_{39}]^2 \left[\dfrac{\frac{[X_6]}{K_1}}{1 + \frac{[X_4][X_6]}{K_1K_2} + \frac{[X_6]}{K_1}}\right] + \\[4mm] + k_{11}k_{14}k_{16}k_{18}k_{19}[X_{39}](k_{-15} + k_{17})\left[\dfrac{\frac{[X_6]}{K_1}}{1 + \frac{[X_4][X_6]}{K_1K_2} + \frac{[X_6]}{K_1}}\right] \end{array}\right\}},$$

or, after some algebraic rearrangements, in the form

$$\frac{1}{[Y']}\frac{d[Y]}{dt} = \frac{k_{14}k_{15}k_{16}k_{17}k_{18}k_{19}[X_{39}][k_{12}[X_4] + k_{11}K_2]}{\left\{\begin{array}{l}[X_4][X_{39}]k_{14}k_{15}k_{17}[k_{16}k_{18}k_{19} + k_{12}k_{18}k_{19} + k_{12}k_{16}] + \\[2mm] + [X_{39}]k_{14}k_{15}k_{16}K_2\left[k_{17}k_{18}k_{19}\left(1 + \frac{K_1}{[X_6]}\right) + k_{11}k_{17} + k_{11}k_{18}k_{19}\right] \\[2mm] + [X_4][k_{12}k_{15}k_{17}k_{18}k_{19}(k_{-14} + k_{16})] + k_{11}k_{14}k_{16}k_{18}k_{19}K_2(k_{-15} + k_{17})\end{array}\right\}}. \tag{7.17}$$

Both of the mathematical formulations, i.e. the system of differential equations (7.10) or a simplified model given by (7.17), supplemented with initial and boundary conditions, provide a detailed description of electrochemical coupling at the myometrial synapse. In practice, the preference for either of them depends on the intended application and is entirely the researcher's choice.

7.2   cAMP-Dependent Pathway

The activated $G\alpha_s$ alpha subunit can bind to and activate the adenylyl cyclase (AC) enzymes. A diversity of ACs – isoforms II–VI and IX have been identified in human myometrium – and their expression at term provides a mechanism for integrating positively or negatively the responses to various transmitters. Types II and IV of ACs have high affinity to $G\beta\gamma$ subunits of inhibitory proteins in the presence of $G\alpha_s$ (Gao and Gilman 1991). Their functionality is affected by phosphorylation with PKC. Types V and VI are inhibited directly by low levels of Ca^{2+} (Yoshimura and Cooper 1992; Premont et al. 1992; Cali et al. 1994). Adenylyl cyclase isofrom III is shown to be up-regulated by a calmodulin-dependent protein kinase in response to the elevation in Ca_i^{2+}, whilst the isoform IX is insensitive to either calcium or $\beta\gamma$-subunits (Mhaouty-Kodja et al. 1997; Hanoune et al. 1997). Thus, it is obvious that the same stimulus may trigger different physiological responses in myometrium depending not only on the type of receptors involved but also on the type of adenylyl cyclase to which they are coupled.

The proposed structure consists of a short amino terminal region and two cytoplasmic domains. The latter are separated by two extremely hydrophobic domains which take the form of six transmembrane helices. The catalytic core of the enzyme consists of a pseudosymmetric heterodimer composed of two highly conserved portions of the cytoplasmic domains. It binds one molecule of $G\alpha_s$ which, in turn, catalyzes the conversion of ATP into cyclic adenosine monophosphate. cAMP influences a wide range of physiological effects including: (1) the increase in Ca^{2+} channel conductance, (2) the activation of protein kinase A (PKA) enzymes, and (3) the cytokine production.

Protein kinase A is a holoenzyme that consists of two regulatory and two catalytic subunits. Binding of cAMP to the two binding sites on the regulatory subunits causes the release of the catalytic subunits and the transfer of ATP terminal phosphates to myosin light chain kinase. The result is a decrease in the affinity of MLCK for the calcium–calmodulin complex and myometrial relaxation. Additionally, PKA may promote relaxation by inhibiting phospholipase C, intracellular Ca^{2+} entry, and by activating BK_{Ca} channels and calcium pumps (Sanborn et al. 1998a, b). PKA activity is controlled entirely by cAMP. Under low levels of cAMP, it remains intact and catalytically inactive.

The level of cAMP is regulated both by the activity of adenylyl cyclase and by phosphodiesterases that degrade it to $5'$-AMP. There is growing experimental evidence that phosphodiesterase 4, specifically PDE4-2B, is increased in pregnancy and facilitates term human myometrial contractions (Leroy et al. 1999; Mehats et al. 2000). Another mechanism that reduces the production of cAMP is the activation of $G\alpha_{i-q/11}$ proteins which directly inhibit adenylyl cyclase through the MAPK signaling cascade. The p38MAPK mechanism has been extensively studied in EP_{3A} receptor (prostaglandin receptor type 3A) stimulation.

The state diagram of the cAMP-dependent pathway as described above is shown in Fig. 7.3.

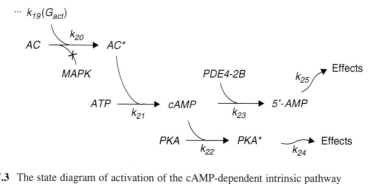

Fig. 7.3 The state diagram of activation of the cAMP-dependent intrinsic pathway

The system of equations for the cAMP-pathway activation is given by

$$d\mathbf{X}/dt = \mathbf{B}\mathbf{X}(t),\qquad(7.18)$$

where $\mathbf{X}(t) = \left(X_j\right)^{\mathrm{T}}\left(j = \overline{21, 26}\right)$ has the components

$$X_{21} = [AC],\quad X_{22} = [ATP],\quad X_{23} = [cAMP],$$
$$X_{24} = [PKA],\quad X_{25} = [PKA^*],\quad X_{26} = [5' - AMP],$$

and the square matrix $\mathbf{B}(b_{ij})(i,j = \overline{6,6})$ has the nonzero elements

$$\begin{aligned}
b_{11} &= -k_{19}k_{20}[X_{18}], & b_{22} &= -k_{21}[X_{27}],\\
b_{32} &= k_{21}[X_{27}], & b_{33} &= -k_{23}[X_{28}]_0,\\
b_{44} &= -k_{22}[X_{23}],\\
b_{54} &= k_{22}[X_{23}], & b_{55} &= -k_{24},\\
b_{63} &= k_{23}[X_{28}]_0, & b_{66} &= -k_{25}.
\end{aligned}$$

The active form of adenylyl cyclase, $X_{27} := AC^*$, is obtained from

$$[X_{27}](t) = [X_{21}]_0 - [X_{21}](t),\qquad(7.19)$$

where $[X_{21}]_0 = \mathrm{const}$ is the initial concentration of the enzyme. We also assume that the level of phosphodiesterase enzyme, $[X_{28}]_0 := \mathrm{PDE}$, remains constant throughout.

The initial conditions provide concentrations of the reacting components

$$\mathbf{X}(0) = \mathbf{X}_0.\qquad(7.20)$$

7.3 PLC Pathway

Activation of oxytocin and/or prostaglandin receptors results in downstream stimulation of the intracellular phospholipase C (PLC)–protein kinase C (PKC) pathway. Four β, two γ, four δ, and ε isoforms of PLC enzymes have been isolated from the human uterus. PLCβ members are triggered by Ca^{2+}, but are differently regulated by G proteins. PLCβ1 and PLCβ4 are sensitive to $G\alpha_{q/11}$, whereas PLCβ2 and PLCβ3 can be activated by $G\alpha_{q/11}$ and $G\beta\gamma$ subunits. Without exception, their activation leads to the break down of inositide-4,5-biphosphate (PIP$_2$) and the generation of second messenger molecules – inositol-1,4,5-triphosphate (IP$_3$) and 1,2-diacylglycerol (DAG) (Omhichi et al. 1995; Stratova and Soloff 1997). IP$_3$ is a highly soluble structure and it quickly diffuses through the cytosol toward the sarcoplasmic reticulum. Here, it binds to R$_{IP3}$ surface receptors and triggers the mobilization of stored Ca^{2+}.

Diacylglycerol consists of two fatty acid chains covalently bonded to a glycerol molecule. Compared to IP$_3$, it is hydrophobic and therefore remains bound to the myometrial plasma membrane. DAG has a number of functions including the activation of protein kinase C, a subfamily of TRPC cation channels, and prostaglandin production. DAG together with acyl-CoA is converted to triacylglycerol by the addition of a third fatty acid to its molecule. The reaction is catalyzed by two distinct isoforms of diglyceride acyltransferases.

Protein kinase C comprises a family of 11 isoenzymes that are divided into three subfamilies – conventional, novel, and atypical – based on their second messenger requirements for activation (Parco et al. 2007). Conventional PKCs (α, β_1, β_2, γ) require free Ca_i^{2+}, DAG, and a phospholipid; novel PKCs (δ, ε, η, θ, μ) need only DAG, and atypical PKCs (ζ, ι, λ) require none of them for activation.

PKCs consist of a variable regulatory and a highly conserved catalytic domain, tethered together by a hinge region. The regulatory domain contains two subregions, namely C1 and C2. The C1 subregion has a binding site for DAG and phorbol esters and C2 acts as a Ca^{2+} sensor and is functional only in PKCs α, β_1, β_2, and γ. Upon their activation, kinases are translocated to the plasma membrane by RACK proteins where they remain active for a long period of time. The effect is attributed to the property of diacyglycerol per se. The enzymes play important roles in several signal transduction cascades. In pregnant myometrium, they phosphorylates MLCK and thus catalyze the contraction response.

The simplified state diagram of the PLC pathway is outlined in Fig. 7.4

The corresponding system of equations is given by

$$dX/dt = CX(t) + C_0, \tag{7.21}$$

where $\mathbf{X}(t) = \left(X_j\right)^{\mathrm{T}} (j = \overline{29, 34})$ has the components

$$X_{29} = [\text{PIP}_2], \quad X_{30} = [\text{IP}_3], \quad X_{31} = [\text{DAG}],$$
$$X_{32} = [\text{DAGT}], \quad X_{33} = [\text{PKC}], \quad X_{34} = [\text{R}_{\text{IP3}}].$$

Fig. 7.4 The state diagram of activation of the PLC-pathway

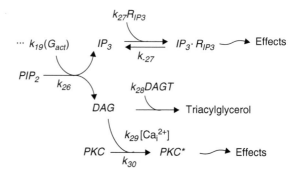

The matrix $\mathbf{C}(c_{ij})(i,j = \overline{6,6})$ has the nonzero elements

$$c_{11} = -k_{19}k_{26}[X_{18}],$$
$$c_{21} = -c_{11}, \qquad\qquad c_{22} = -k_{27},$$
$$c_{31} = c_{21}, \qquad\qquad c_{33} = -k_{28}[X_{32}] - k_{29}k_{30}[X_1][X_{33}],$$
$$c_{44} = -k_{28}[X_{31}],$$
$$c_{55} = -k_{29}k_{30}[X_1][X_{31}],$$
$$c_{66} = -k_{27}[X_{30}] - k_{-27},$$

and the vector-column is: $\mathbf{C_0} = (0, k_{-27}[X_{37}], 0, 0, 0, k_{-27}[X_{37}])^T$.

Concentrations of the active form of protein kinase C, $X_{35} := $ PKC*, and the activated IP$_3$–R$_{IP3}$ complex, $X_{36} := $ IP$_3$ · R$_{IP3}$, can be obtained from the algebraic relations

$$
\begin{aligned}
[X_{35}](t) &= [X_{38}]_0 - [X_{33}](t) \\
[X_{36}](t) &= [X_{37}]_0 - [X_{34}](t).
\end{aligned}
\tag{7.22}
$$

Here $[X_{37}]_0, [X_{38}]_0$ are the initial concentrations of the IP$_3$-receptor on the endoplasmic reticulum and protein kinase C enzyme, respectively.

The dynamics of intracellular calcium release from the stores is described by Miftahof et al. (2009)

$$
\frac{d[X_1]}{dt} = \widetilde{k}_0 \left(\widetilde{k}_1 - \widetilde{k}_2[X_{36}]^3[X_1] - [Ca^{2+}_{SR}] \right) - \frac{\widetilde{k}_3[X_1]}{[X_1]^2 - \widetilde{k}_4},
\tag{7.23}
$$

where $\widetilde{k}_i(i = 0, 4)$ are the kinetic parameters related to the release of sarcoplasmic Ca$^{2+}_{SR}$.

Provided that initial concentrations of reactive substrates are known, the system of (7.21)–(7.23) models the PLC pathway dynamics.

7.4   Co-transmission in Myometrium

Immunohistochemical research has convincingly shown that individual cells can colocalize more than two biological substances which possess functional properties and may act as transmitters or modulators (Nusbaum et al. 2001; Merighi 2002; Burnstock 2004; Trudeau and Gutiérrez 2007; Teschemacher and Christopher 2008). Such coexistence suggests that a cell may simultaneously release more than one substrate and that co-transmission may occur. The phenomenon is ubiquitous and is therefore considered the rule rather than the exception. Indeed, co-release of classical neurotransmitters – glutamate, γ-amino butyric acid, acetylcholine, dopamine – and peptides – ATP, neurotrophic factor, nitric oxide, endogenous cannabinoids – among others – has been well established throughout different parts of the body. The process is highly regulated in response to various physiological, chemical, and pathological signals. However, the exact mechanisms and how it actually occurs in the human uterus remains an open question that has not been addressed in details yet.

Functional co-transmission implies that various postsynaptic receptors exist in the vicinity of the presynaptic terminal or an adjacent cell. In the context of such a configuration, one transmitter/modulator could affect the action of another. For example, it may shunt and modify the time course of excitation/inhibition signals, provide presynaptic neuromodulation and/or synapse-specific adaptation, control differentially modulation of neuroeffector circuit, activate synergistically multiple receptors of the same or different postjunctional cells. It also adds a safety factor to the communication process and thus compensates for activity- or pathology-dependent alterations in postsynaptic receptor subunits. These mechanisms appear to be extremely useful, because cell-to-cell signaling is not fully constrained by synaptic wiring and synapse independence.

The quantitative features of integrative physiological phenomena in the pregnant human uterus are defined by contributions from the myriad of interconnecting cross-talk intracellular signaling pathways. The combined state diagram of ligand–receptor binding, activation of second messenger systems and exertion of electro-mechanical effects is given in Fig. 7.5, and is summarized in Table 7.1. It is apparent that elements of one pathway cross-regulate and share components of another pathway in the transduction process.

The corresponding governing system of equations is given by

$$dX/dt = DX(t) + C_0, \qquad (7.24)$$

where

$$
\mathbf{D}(m \times n) =
\begin{pmatrix}
\begin{matrix} a_{ij} \\ (i,j=1,20) \end{matrix} & 0 & 0 \\
0 & \begin{matrix} b_{kl} \\ (k,l=1,6) \end{matrix} & 0 \\
0 & 0 & c_{kl}
\end{pmatrix},
\quad
\mathbf{X} =
\begin{pmatrix} X_1 \\ \vdots \\ X_{26} \\ X_{29} \\ \vdots \\ X_{34} \end{pmatrix},
\quad
\mathbf{C_0} =
\begin{pmatrix} 0 \\ \vdots \\ 0 \\ C_{30} \\ \vdots \\ C_{34} \end{pmatrix}.
$$

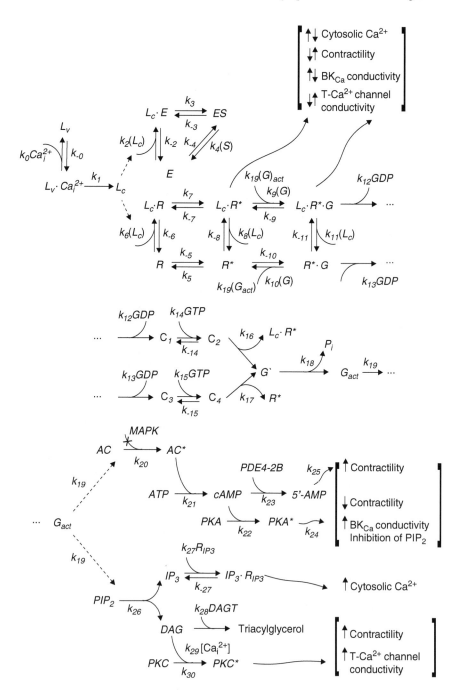

Fig. 7.5 Proposed generalized scheme of neurohormonal transmission and regulation in human myometrium

Table 7.1 Selected cross-talk signaling components in the human uterus

Transmitter/modulator	Receptor type	$G_{(act)}$-protein	Pathway	Ion channel
ACh	μ_2	$G\alpha_i$	$\downarrow AC^*$	
AD	α_1	$G\alpha_{q/11}$	$\uparrow PLC$	
	α_2	$G\alpha_i$	$\downarrow AC^*$	
	β_1, β_3	$G\alpha_s$	$\uparrow AC^*$	$\uparrow BK_{Ca}$
	β_2	$G\alpha_s \,\&\, G\alpha_i$	$\uparrow AC^*$	$\uparrow BK_{Ca}$
PrF$_{2\alpha}$	FP	$G\alpha_{q/11}$	$\uparrow PLC$	$\uparrow L - Ca_i^{2+}$
PrE$_2$	EP$_{1,3D}$	–	–	$\uparrow L - Ca_i^{2+}$
	EP$_{3A}$	$G\alpha_i$	$\downarrow AC^*$	
	EP$_2$	$G\alpha_{q/11}$	$\uparrow PLC$	
OT	OTR	$G\alpha_{q/11}$	$\uparrow PLC$	$\uparrow T - Ca_i^{2+}$
PR	mPR	$G\alpha_i$	$\downarrow AC^*$	

The elements of $\mathbf{D}, \mathbf{X}$, and $\mathbf{C_0}$ are defined by (7.10), (7.18), and (7.21). Algebraic relationships (7.19) and (7.22) are used to calculate current concentrations of the reactants X_{27}, X_{35} and X_{36}, while the values of $X_{28,37-41}$ refer to initial concentrations of substrates and are defined a priori.

Assuming the level of separate or conjoint activation of final components, i.e. $L_c \cdot R^*, 5' - AMP, PKA^*, PKC^*$ and $IP_3 \cdot R_{IP3}$, is taken as a measure of pathway output, responses of the myometrium to stimulatory signals may be:

1. Depolarization of the cell membrane, which is described by (7.11)
2. Increase in permeability of ion channels

$$g_p(t) = g_p[X_q]/[X_q]_{max}, \qquad (7.25)$$

where the channel selectivity (p) depends on the transmitter and the receptor type (q) involved: $(p, q) \propto (BK_{Ca}, PKA^*), (T - Ca_i^{2+}, OTR^*), (L - Ca_i^{2+}, PrF_{2\alpha})$, $(L - Ca_i^{2+}, PrE_2)$,

3. Augmentation of contraction/relaxation, which in one-dimensional case, T^a, is given by (3.6), and in case of in-plane forces, σ_{kl}^2, by (6.19).

7.4.1 Co-transmission by Acetylcholine and Oxytocin

Consider the effect of co-transmission by acetylcholine and oxytocin on the dynamics of the myometrial fasciculus. The combined system of equations (3.4)–(3.18), (7.1)–(7.5), (7.24), (7.11) and (7.25), for $p \propto T - Ca_i^{2+}$, $q \propto OTR^*$, provides the mathematical formulation of the problem and describes:

1. The excitation and release of the transmitters/modulators
2. The electrochemical coupling at the synapse
3. Intracellular processes of transduction, and
4. Myometrial responses.

Note that the role of the stimulatory signal, V^s, in the boundary condition (3.17) should be replaced by the synaptic potential, V_{syn}, given by (7.11).

Conjoint effects of ACh and OT were studied numerically. The amount of neurotransmitter released depended on the strength of depolarization of the nerve terminal, V^f. Effects of different concentrations of oxytocin were modeled by varying the conductivity parameter for the fast Ca^{2+} channel, g_{Ca}^f.

Results of simulations show that the depolarization of the presynaptic membrane activates a short-term influx of extracellular calcium ions into the terminal through voltage-gated Ca^{2+} channels. The concentration of cytosolic Ca_i^{2+} quickly rises and reaches the maximum -19.4 μM. Some of the ions are immediately absorbed by the buffer system, others diffuse toward vesicles. They bind to the active centers and initiate the release of acetylcholine. During the whole cycle, about 10% of stored vesicular neurotransmitter is released.

Part of the postsynaptic acetylcholine undergoes degradation by acetylcholine esterase: max $[ACh-E] = 0.47$ μM is formed. The complex quickly dissociates into the enzyme and choline, S, which is reabsorbed and drawn into a new cycle of ACh synthesis.

The diffused fraction of ACh in the synaptic cleft equals 3.2 mM. The major part of the transmitter enters the postsynaptic membrane where it reacts with the surface receptors to form the ACh–R complex: $[ACh-R] = 0.123$ μM. Their transformation to the active state triggers the generation of the excitatory postsynaptic potential, V_{syn}. It increases as a step function and reaches 87 mV in 0.25 ms.

The myometrial fasciculus generates high amplitude, $V \simeq 30$ mV, and frequency, $v = 3.2$ Hz, action potentials on the crests of slow waves (Fig. 7.6). There is a concurrent increase in the level of intracellular calcium. The maximum concentration $\left[Ca_i^{2+}\right] = 0.49$ μM is recorded. As a result, the myofiber produces regular contractions of the strength 15.8 mN/cm, and a duration of 2.5 min.

Conjoint release of ACh and OT into the system causes a further rise in the amplitude of spikes, $V \simeq 38$ mV, and the concentration of intracellular calcium: $\left[Ca_i^{2+}\right] = 0.51$ μM. There is a slight increase in the intensity of contractions. The maximum active force, $T^a = 16.1$ mN/cmcm is produced, of an average duration of 2 min.

Washout of ACh abolishes high amplitude spiking activity. Oxytocin acting alone induces irregular action potentials of amplitude 7–10 mV. Remarkably, there are no significant changes in the transient and concentration of free cytosolic calcium. The fasciculus continues to contract with the strength of 16 mN/cm.

7.4.2   Co-transmission by Acetylcholine and Adrenaline

Consider the effect of co-transmission by ACh and adrenaline (AD) on myoelectrical activity of the fasciculus. The system of equations including (3.4)–(3.18), (7.1)–(7.5), (7.24) and (7.11) provides the mathematical formulation of the problem. Results of numerical simulations are discussed below.

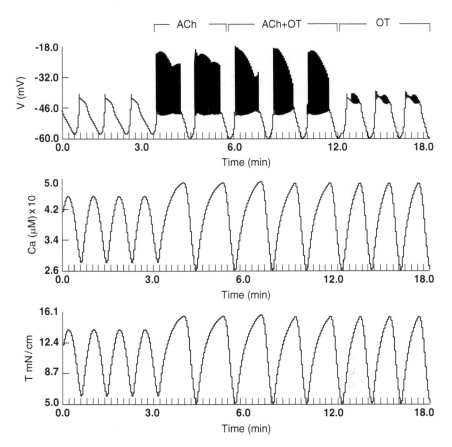

Fig. 7.6 Selective and conjoint effects of acetylcholine (ACh) and oxytocine (OT) on electrome-chanical activity of the fasciculus

Excitation of the nerve terminal triggers the release of vesicular AD. The maximum 86.3 μM of the free fraction of neurotransmitter is released in the synaptic cleft. AD_c quickly diffused toward the postsynaptic membrane where part of it, $[AD_c] = 35.2$ μM, binds to $\beta_{2,3}$-adrenoceptors while another part is utilized by the re-uptake mechanisms. The total of 8.53 μM of the $AD_c \cdot R$ complex is being formed. This causes hyperpolarization of the myometrium to -78.8 mV.

Subsequent release of AD into the pre-excited by ACh myometrial synapse has a detrimental effect on its electrical and mechanical activity (Fig. 7.7). The amplitude of slow wave oscillations reduces to 17 mV while their frequency incre-ases twofold. The rate of spiking and their amplitude decrease and the myomet-rium becomes hyperpolarized, $V = -67$ mV. These changes reflect the fall in the concentration of free cytosolic calcium, $\left[Ca_i^{2+}\right] = 0.27$ μM, and contractility, max $T^a = 6.0$ mN/cm.

Adrenaline alone in low concentration reduces excitability of the fasciculus and suppresses the intensity of active forces. No action potentials are being generated

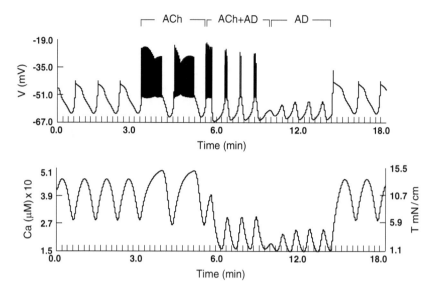

Fig. 7.7 Selective and conjoint effects of acetylcholine (ACh) and adrenaline (AD) on electro-mechanical activity of the fasciculus

and $T^a = 4.3$ mN/cm develops. Further increase in AD causes complete relaxation of the muscle.

7.4.3 Co-transmission by Oxytocin and Adrenaline

Results of numerical simulations of conjoint action of adrenaline and oxytocin on the myometrial fasciculus are shown in Fig. 7.8. Application of AD to the system that is already exposed to OT has a significant hyperpolarizing effect. The resting membrane potential level falls to −67 mV and the amplitude of slow waves decreases to 10 mV. These changes have a negative effect on the dynamics of Ca^{2+}. Only 0.17 μM of free cytosolic calcium ions are recorded. They sustain weak contractions of a maximum strength of 2.0 mN/cm. Further washout of OT from the system abolishes spontaneous contractions and relaxes the myometrium.

7.4.4 Co-transmission by Oxytocin and Prostaglandins

Addition of prostaglandins to the myometrial fasciculus pretreated with OT causes a short burst of high frequency spikes of various amplitudes. The intracellular calcium concentration reaches 0.53 μM and coincides with the maximum active force development $T^a = 17$ mN/cm of a duration of ~3 min (Fig. 7.9).

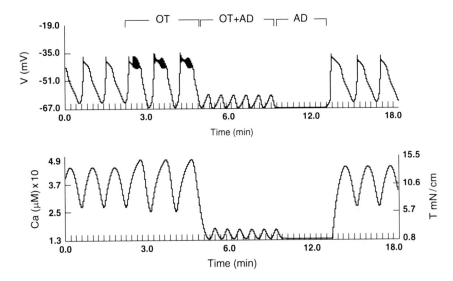

Fig. 7.8 Selective and conjoint effects of oxytocin (OT) and adrenaline (AD) on electromechanical activity of the fasciculus

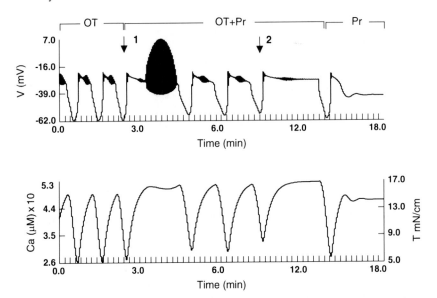

Fig. 7.9 Selective and conjoint effects of oxytocin (OT) and prostaglandin (Pr) on electromechanical activity of the fasciculus

Further increase in the level of prostaglandins stably depolarizes the membrane at -20 mV and causes a tonic-type contraction of the fasciculus, $T^a = 17.2$ mN/cm.

Prostaglandins acting alone abolish slow waves and depolarize myometrium at $V = -20$ mV. The cytosolic calcium content falls to 44 μM as well as the active force of contraction, $T^a = 14.4$ mN/cm.

Exercises

1. What is the evidence that supports the model of a myometrial synapse as an open system?
2. State the basic assumptions of the Michaelis–Menten kinetics. Discuss their validity in the model of neurotransmission in the myometrial synapse.
3. What are the sources of limitations for the Michaelis–Menten kinetics?
4. The Michaelis equilibrium constant, K^*, is an approximation of the affinity of the enzyme for the substrate based on the rate constants within the reaction. It is numerically equivalent to the substrate concentration at which the rate of conversion is half of the maximum reaction rate. How can these constants be determined experimentally? (*Hint*: study a Lineweaver–Burk plot).
5. Provide experimental justification for the boundary condition $V^f(L, t) = 0$ (7.8).
6. Formulate a mathematical problem of the release of prostaglandins $F_{2\alpha}$ and E_2 mediated by the multidrug resistance-associated protein 4.
7. Formulate a mathematical problem of downregulation of type V and VI adenylyl cyclase isoforms activity by intracellular Ca_i^{2+} turnover (*Hint*: consider the role of intracellular stores in Ca_i^{2+} dynamics).
8. Modify the mathematical formulation (Fig. 7.3) of cAMP-dependent pathway activation that would include the conversion of regulatory and catalytic subunits of the PKA holoenzyme.
9. Modify the mathematical formulation (Fig. 7.4) of PLC pathway activation to detail the function of C1 and C2 regulatory subunits of the PKC enzyme.
10. There is experimental evidence to support the view that extracellular adenosine 5′-triphosphate regulates myometrial activity via P2X and P2Y receptors. Construct a mathematical model of purinergic neurotransmission at the myometrial synapse.
11. Simulate the effect of the cyclic guanosine monophosphate phosphodiesterase 5-specific inhibitor, sildenafil citrate (Viagra), on contractility of myofiber (*Hint*: sildenafil citrate exerts its effect independently of cyclic guanosine monophosphate but acting directly at potassium channels).
12. It has been shown that 5-hydroxytryptamine (serotonin) regulates human myometrial activity. Find the receptor subtypes involved in neurotransmission. Construct a mathematical model of a serotonergic myometrial synapse. Simulate the effects of alpha-methyl-5-hydroxytryptamine and ketanserin on myofiber.

Chapter 8
Pharmacology of Myometrial Contractility

(The scientist) ... if dissatisfied with any of his work, even if it be near the very foundations, can replace that part without damage to the remainder.

G.H. Lewis

8.1 Classes of Drugs

The majority of pharmacological agents used in clinical practice act to alter the processes responsible for transmission, by facilitating or inhibiting: (1) release, (2) enzymatic degradation of the neurotransmitter or modulator, (3) function of specific postsynaptic receptors, (4) second messenger system, or (5) intracellular regulatory pathways. For example, N-type calcium ion channel blockers – derivatives of ω-conopeptides – interfere with the dynamics of cytosolic Ca_i^{2+} in the presynaptic nerve terminal. The decreased Ca_i^{2+} concentration prevents activation of calmodulin protein and movement of vesicles containing neurotransmitters toward presynaptic membrane. Chemical agents that facilitate cholinergic and adrenergic neurotransmission can inhibit true and pseudo-acetylcholinesterase, monoamine oxidase, and catechol-O-methyltransferase enzymes in the synaptic cleft.

There are more than 20 families of receptors that are present in the plasma membrane, altogether representing over 1,000 proteins of the receptorome (Strachan et al. 2006). Transmembrane and intracellular receptors are being used as drug targets and they have a wide array of potential ligands. However, to date only a small percentage of the receptorome has been characterized. Abilities of a ligand to react with a receptor depend on its specificity, affinity, and efficacy. Selectivity is determined by the chemical structure and is related to physicochemical association of the drug with a recognition (orthosteric) site on the receptor. The probability at which the ligand occupies the recognition site is referred to as affinity, and the degree at which the drug produces the physiological effect is defined as efficacy. Because multiple receptors are expressed on human myometrium, it is evident that disparate ligands acting alone or conjointly may elicit similar cellular responses.

R.N. Miftahof and H.G. Nam, *Biomechanics of the Gravid Human Uterus*,
DOI 10.1007/978-3-642-21473-8_8, © Springer-Verlag Berlin Heidelberg 2011

A given receptor may contain one or more binding sites for various ligands and can be linked to different second messenger systems. The interaction of a drug with the site that is topographically distinct from the orthosteric site is called allosteric (Monod et al. 1965). The essential features of an allosteric drug–receptor interaction are: (1) the binding sites are not overlapping, (2) interactions are reciprocal in nature, and (3) the effect of an allosteric modulator could be either positive or negative with respect to association and/or function of the orthosteric ligand. Thus, binding of a drug to the receptor changes its conformational state from the original tense to the relaxed form and hence, either facilitates or inhibits linking of the endogenous transmitter.

Drugs that act at receptors are broadly divided into agonists and antagonists. Ligands that interact with the orthosteric site of a receptor and trigger the maximum response are called full agonists. Related to them structurally are partial agonists, however, with lower biological efficacy. They are regarded as ligands with both agonistic and antagonistic effects, i.e. in the case of conjoint application of a full and partial agonist, the latter competes for receptor association and causes a net decrease in its activation (Kenakin 2004). In practice, partial agonists either induce or blunt a physiological effect, depending on whether inadequate or excessive amount of endogenous transmitter is present, respectively.

Receptors that exhibit intrinsic basal activity and may initiate biological effects in the absence of a bound ligand are called constitutively active. Their function is blocked by application of inverse agonists – drugs that not only inhibit the association of an agonist with the receptor but also interfere with its activity. This pharmacokinetic characteristic distinguishes them from true competitive antagonist (see below). Many drugs that have been previously classified as antagonists are being reclassified as inverse agonists.

A class of drugs that have selectivity and affinity, but no efficacy for their cognate receptor are called antagonists. Antagonists that interact reversibly at the active site are known as competitive. Once bound, they block further association of an agonist with the receptor and thus prevent the development of a biological response. Ligands that react allosterically are called noncompetitive. They stop conformational changes in the receptor necessary for its activation. A subtype of noncompetitive ligands that require agonist–receptor binding prior to their association with a separate allosteric site is called uncompetitive. Their characteristic property is related to the effective block of higher, rather than lower, agonist concentrations.

Ligands that affect second messenger system function are classified according to the enzyme they act on. Thus, there are competitive selective and nonselective cyclic adenosine monophosphate, protein kinase A and C, phosphodiesterase, diacylglycerol, Ca^{2+}-adenosine triphosphatase, etc., activators and inhibitors. Because a single enzyme is often involved in multiple regulatory pathways, drugs of this category have a narrow, therapeutic index and many side effects. Despite their pharmacokinetic and clinical limitations, they are widely used in laboratory research and represent a new approach in the management of inflammation-induced preterm delivery.

All drugs, depending on the stability of a drug–acceptor complex that is being formed, show reversible or irreversible interaction. Reversible ligands have strong chemical affinity to a natural transmitter or modulator and normally form an unstable complex which quickly dissociates into a drug and a "receptor." In contrast, irreversible drugs are often chemically unrelated to the endogenous transmitter and covalently bind to the target creating a stable complex.

8.2 Current Therapies of Myometrial Dysfunction

The control of myometrial excitability has important therapeutic implications since myometrial dysfunction may lead to preterm labor or dystocia. These conditions bring numerous complications, both for mother and fetus, and are a major cause of perinatal mortality and morbidity. The gestational age gaining a few weeks, or even a few days, increases the chances of survival and lowers considerably the risk of handicap for the newborn. In practice, the prolongation of pregnancy, when indicated, is achieved by using a number of tocolytic (myometrial relaxing) agents. Based on their pharmacological mechanisms of action they are classified as calcium channel antagonists (nifedipine), cytosolic calcium lowering drug (magnesium sulfate), β_2-adrenoceptor agonists (ritodrine, terbutaline, isoxsuprine), nonsteroidal anti-inflammatory drugs (indomethacin), and myometrial relaxants (17α hydroxy-progesterone caproate, nitroglycerine) (Bernal and TambyRaja 2000). Recent research in that area has been focused on the discovery of oxytocin antagonists and PDE4 inhibitors (Slattery and Morrison 2002; Romero et al. 2000; Mehats et al. 2000, 2007; Oger et al. 2004; Serradeil-Le Gal et al. 2004).

Although approximately 1.3 million women annually worldwide benefit from existing treatment, for the last two decades no new drugs targeting specifically uterine activity have been developed. All currently existing recommendations suggest new combinations of well-known drugs, changes in dosage, and ways of administration. Much of the concern over the use of these therapies is related to the associated adverse effects, including limited capacity to delay labor, and no proven advantage for neonatal outcome and fatality. Therefore, the development of specific and effective tocolytic agents remains an urgent goal.

Successful drug discovery requires deep understanding of the mechanisms of diseases, the full biological context of the drug target, biochemical mechanisms of drug action, and should involve a multilevel conceptual framework. Systems biology approach, as a systematic interrogation of biological processes within the physiological milieu in which they function, provides a new paradigm to study the combined behavior of interacting components through the integration of experimental, mathematical, and computational methods.

The development of drugs for clinical use calls for standard test models in which a large number of substances can be analyzed simultaneously to determine whether, and how, drugs contribute to the overall pharmaco-physiological response. Mathematical modeling and simulations in the form of virtual laboratories provide a platform for bridging gaps in our understanding of intricate mechanisms and generation of new hypotheses that are not realizable within a real laboratory. Verified experimentally, they serve as predictors in the iterative process of identification, validation and evaluation of drug targets, and provide a valuable insight into questions of drug efficacy and safety.

8.3   Model of Competitive Antagonist Action

Selective competitive reversible oxytocin receptor (OTR) antagonists – atosiban (approved) and barusiban (under clinical development) – are designed to manage preterm labor. Both drugs demonstrate high efficacy and rapid onset of action (0.5–1.5 h). The inhibitory effect of them can be reversed within 1.5–2.5 h after application by high doses of oxytocin infusion (Reinheimer et al. 2005; Reinheimer 2007; Richter et al. 2005; Wex et al. 2009). Studies in human liver microsomes have shown that atosiban is not metabolized by the cytochrome P450-enzyme system. The drug is excreted either unchanged or by sequential hydrolysis of the C-amino acid terminal in urine and feces. Atosiban is frequently administered together with labetalol, a nonselective α and β-adrenoceptor antagonist, to treat pregnancy-induced hypertension and associated pre-eclampsia. Although studies conducted on healthy pregnant women revealed a lack of pharmacokinetic interactions between the two drugs and no serious adverse effects, prognostic assessment for potential complications in various clinical scenarios remains an unresolved problem (Rasmussen et al. 2005).

Let L_{At} be the competitive reversible antagonist. A part of the state diagram of signal transmission in the myometrium that describes ligand action is shown in Fig. 8.1, while other reactions in the general cycle remain unchanged (Fig. 7.5).

Assuming that the reactions of association/dissociation of the antagonists (L_{At}) with the receptor (R) and the drug–receptor complex ($L_{At} \cdot R$) formation satisfy the Michaelis–Menten kinetics, the governing system of (7.24) is

$$dX_{At}/dt = D_{At}X_{At}(t) + C_{0,At}. \tag{8.1}$$

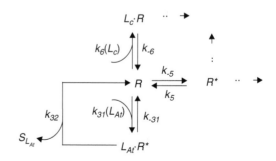

Fig. 8.1 The state diagram of signal transduction in myometrium in presence of a reversible antagonist

The matrix $\mathbf{D}_{At}$ is the extension of the matrix $\mathbf{D}$ (7.10)

$$
\mathbf{D}_{At} = \begin{pmatrix} a_{ij} & \cdots & & 0 & 0 \\ & & b_{kl} & 0 & 0 \\ 0 & \cdots & 0 & c_{kl} & 0 \\ & & & d_{35,35} & d_{35,36} \\ 0 & \cdots & & d_{36,35} & d_{36,36} \end{pmatrix}, \quad
\mathbf{X}_{At}(t) = \begin{pmatrix} X_1 \\ \vdots \\ X \\ X_{29} \\ \vdots \\ X_{34} \\ X_{42} \\ X_{43} \end{pmatrix}, \quad
\mathbf{C}_{0,At} = \begin{pmatrix} 0 \\ \vdots \\ 0 \\ C_{30} \\ \vdots \\ C_{34} \\ 0 \\ 0 \end{pmatrix},
$$

where the modified elements and new elements are given by

$$
\begin{aligned}
a_{5,21} &= -k_{31}[X_5], & a_{5,22} &= k_{-31} + k_{32}, \\
d_{35,35} &= -k_{31}[X_5], & d_{35,36} &= k_{-31}, \\
d_{36,35} &= k_{31}[X_5], & d_{36,36} &= -(k_{-31} + k_{32}).
\end{aligned}
$$

New components of the vector $\mathbf{X}_{At}$ are: $X_{42} := L_{At}$, $X_{43} := L_{At} \cdot R$. To close the system, it should be complemented by initial values for L_{At} and $L_{At} \cdot R$.

In case of partial chemical equilibrium, which could be achieved after prolonged treatment of myometrium with the antagonist, the dynamics of the receptor, drug and drug–receptor complex conversions can be described by,

$$
\begin{aligned}
\frac{d[X_{5(42)}]}{dt} &= -k_{31}[X_5][\overline{X}_{42}] + k_{-31}[\overline{X}_{43}] \\
\frac{d[X_{43}]}{dt} &= k_{31}[X_5][\overline{X}_{42}] - k_{-31}[\overline{X}_{43}].
\end{aligned} \tag{8.2}
$$

Here $[\overline{X}_{42}]$, $[\overline{X}_{43}]$ are equilibrium concentrations of the drug and the bounded complex, respectively. Summation of equations for $[X_5], [X_{43}]$ yields

$$
\frac{d[X_5]}{dt} + \frac{d[X_{43}]}{dt} = 0 \text{ or } [X_5](t) + [\overline{X}_{43}](t) = \text{const}, \tag{8.3}
$$

from where letting the concentration of total available receptors $[X_5]_0 = $ constant, we get

$$
[X_5](t) = [X_5]_0 - [\overline{X}_{43}](t). \tag{8.4}
$$

Since $d(L_{At} \cdot R)/dt = 0$, substituting (8.4) into the second equation of (8.2) and after simple algebra, we obtain

$$[\overline{X}_{43}](t) = \frac{K^*[X_5]_0}{K^* + [\overline{X}_{42}]^{-1}}. \tag{8.5}$$

Here $K^* = k_{31}/k_{-31}$ is the Michaelis–Menten equilibrium constant.

Finally, the dynamics of receptors in presence of a competitive antagonist and a corresponding endogenous transmitter (L_c), e.g. oxytocin, acetylcholine, adrenaline, is given by

$$[X_5](t) = [X_5]_0 - [\overline{X}_{43}](t) - [X_6](t) - [\overline{X}_8](t). \tag{8.6}$$

Here $[X_6](t)$ is the concentration of constitutively active receptors, and $[\overline{X}_8](t)$ is the equilibrium concentration of the $L_c \cdot R$ - complex.

Substitution of (8.6) into (8.1) allows some simplifications in the governing system of equations. We have left it as an exercise to the diligent reader.

8.4 Model of Allosteric Interaction

Two broad conceptual views underlie the majority of studies of allosterism. The first, that was developed initially in the field of enzymology, is based on the assumption that proteins possess more than one binding site that can react successively with more than one ligand. The second considers allosterism as the ability of receptors to undergo changes that eventually yield an alteration in affinity of the orthosteric sites for endogenous transmitters (Monod et al. 1965; Koshland et al. 1966).

Respectively, two types of mathematical models, i.e. concerted and sequential, along with their various expansions and modifications, have been proposed to simulate cooperative binding. The concerted model assumes that:

1. Enzyme (receptor) subunits in equilibrium attain the identical – tensed or relaxed – conformation
2. It is affected by an allosteric effector
3. A conformational change in one subunit is conferred equally to all other subunits.

In contrast, the sequential model does not require the satisfaction of conditions (1), (3) but instead dictates an induced fit binding of the ligand with subsequent molding of the target, instead.

Consider a modified part of the general state diagram of allosteric ligand–receptor interaction (Fig. 8.2). Positive noncompetitive allosteric mechanism assumes binding of the ligand L_{All} to the receptor in the R^T conformation. The $L_{All} \cdot R^R$ - complex further associates with the endogenous transmitter L_c and produces the active complex – $L_c \cdot R^R \cdot L_{All}$. In case on uncompetitive positive allosteric mechanism, though, the transmitter L_c binds first to the R^T - receptor, changes its configuration to $L_c \cdot R^R$ form and only then the ligand L_{All} occupies the allosteric site. The $L_c \cdot R^R \cdot L_{All}$ - complex reacts with the G-protein system and enters the cascade of chemical transformations as described above (Fig. 7.5).

Fig. 8.2 The state diagram of allosteric ligand–receptor interaction

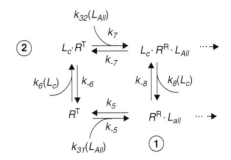

Comparison of the unperturbed and current schemes gives: $X_5 := R^T$, $X_6 := L_{All} \cdot R^R$, $X_7 := L_c \cdot R^R \cdot L_{All}$, $X_8 := L_c \cdot R^T$. Assuming that all reactions satisfy the Michaelis–Menten kinetics, the system of equations for allosteric interaction is given by

$$d\mathbf{X}_{All}/dt = \mathbf{D}_{All}\mathbf{X}_{All}(t) + \mathbf{C}_{0,All}. \tag{8.7}$$

Here $\mathbf{X}_{All}(t) = (X_1, ..., X_{34}, X_{44})^T$, $\mathbf{C}_{0,At} = (0, ...C_{30}, 0, ...0, C_{34}, 0)^T$, $X_{44} := L_{All}$. The matrix $\mathbf{D}_{All}$ contains the modified matrix $\mathbf{D}$ (7.24) and the additional new elements

$$a_{4,35} = k_{-6}[X_8] + k_{31}[X_5],$$
$$a_{5,35} = k_{-6}[X_8] - k_{31}[X_5],$$
$$a_{6,35} = k_{31}[X_5],$$
$$a_{7,35} = k_{32}[X_8],$$
$$a_{8,35} = -(k_{-6} + k_{32})[X_8],$$
$$d_{35,4} = k_5[X_6] + k_6[X_5], \; d_{35,7} = k_{-7}, \; d_{35,35} = -k_{31}[X_5](k_{-6} + k_{32})[X_8].$$

In contrast to competitive antagonists which cause a theoretically limitless rightward shift of the dose–occupancy and dose–effect curves for endogenous transmitter, allosteric ligands attain a limit which is defined by the binding factor. Thus, allosteric agonists, applied conjointly with agonists and endogenous transmitters, can enhance their spatial and temporal selectivity at given receptors and guarantee a required level of safety.

Recently, a novel precursor of a drug, PDC113.824, an allosteric modulator of prostaglandin $F_{2\alpha}$, has been investigated for its tocolytic properties and prevention of preterm labor. Experiments showed that the compound blocked $PGF_{2\alpha}$-mediated $G\alpha_{1/2}$-dependent activation of the Rho/ROCK signaling pathways, actin remodeling, and contraction of human myometrial cells. Additionally, it also showed positive modulation of PKC and MAP kinase signaling. This "bias" in receptor-dependent transmission was explained by an increase in $PGF_{2\alpha}$ receptor coupling to $G\alpha_q$ second messenger system at the expense of $G\alpha_{1/2}$ proteins (Goupil et al. 2010).

8.5   Allosteric Modulation of Competitive Agonist/Antagonist Action

One of the intriguing pharmacological properties of allosteric drugs is their potential ability to alter selectivity, affinity, and efficacy of bound and nonbound competitive agonists/antagonists by enhancing or inhibiting their cooperativity at receptor sites. During the last decade, the effect of different allosteric compounds has been studied extensively in vivo and in vitro. For example, it was found that gallamine diminishes the affinity of bound acetylcholine and inhibits its negative ionotropic and chronotropic effects in the myocardium, while alcuronium exerts positive allosteric modulation on the affinity of ACh (Stockton et al. 1983; Tucek et al. 1990). Radioligand binding studies with the human adenosine A_1 receptor revealed the diverse regulatory effects of the allosteric modulator (PD81,723) on the affinity of a partial agonist (LUF5831), a full agonist (N^6-cyclopentyl-adenosine (CPA)), and an inverse agonist/antagonist (8-cyclopentyl-1,3-dipropylxanthine – DPCPX), for the receptor. Results demonstrated that it increased the affinity of CPA, slightly decreased the affinity of LUF5831, and significantly reduced the affinity of DPCPX (Heitman et al. 2006). Estrogen has been shown to modulate allosterically the effects of oxytocin through the activation of transcription of reporter plasmids, the recruitment of specific proteins and conformational changes in the estrogen receptor (Wood et al. 2001). Thus, therapeutically, allosteric ligands are capable of modifying signals carried by the exogenous and/or endogenous ligands in the system.

Let the myometrium under consideration be exposed simultaneously to a competitive agonist (antagonist), $L_{Ag(ant)}$, allosteric ligand, L_{All}, and the endogenous transmitter, L_c. The proposed state diagram of their interactions is shown in Fig. 8.3.

It combines noncompetitive and uncompetitive allosteric mechanisms of action which involve binding of L_{All} to the receptor R^T, formation of the $L_{All} \cdot R^R$-complex followed by binding of the agonist (antagonist) $L_{Ag(ant)}$ and the transmitter L_c to it, and an inverse sequence, i.e. binding of $L_{Ag(ant)}$ and L_c to the receptor R^T first with the subsequent addition of the ligand L_{All}. As a result of both processes, the $L_c \cdot L_{Ag(ant)} \cdot R^R \cdot L_{All}$ active (inactive) complex is produced.

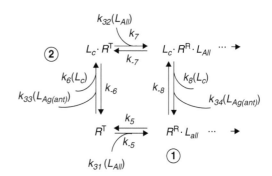

Fig. 8.3 The state diagram of the effects of a competitive agonist/antagonist and allosteric ligand on the neuro-hormonal transmission in myometrium

The governing system of equations for positive noncompetitive allosteric modulation of the agonist (antagonist) action is

$$dX_{Am}/dt = D_{Am}X_{Am}(t) + C_{0,Am}. \tag{8.8}$$

The vector of reacting components is $X_{Am}(t) = (X_1, ..., X_{34}, X_{42}, ..., X_{46})^T$, where $X_5 := R^T$, $X_6 := L_{All} \cdot R^R$, $X_7 := L_c \cdot R^R \cdot L_{All} \cdot L_{Ag}$, $X_8 := L_c \cdot R^T \cdot L_{Ag}$, $X_{42} := L_{Ag}$, $X_{43} := L_{Ag} \cdot R^T$, $X_{44} := L_{All}$, $X_{45} := L_{All} \cdot R^R \cdot L_{Ag}$, $X_{46} := L_{All} \cdot R^R$. L_c. The meaning of other components is as described in (7.24). The vector of constant concentrations of substrates is $C_{0,Am} = (0, ...C_{30}, 0, ...0, C_{34}, 0, ...0)^T$.

The matrix D_{Am} is obtained from the general matrix D where the following elements are adjusted and new elements are introduced

$a_{44} = k_2([E_0] - [X_8]) + k_6[X_5] + (k_{-8} + k_5 + k_6)[X_6] + k_{-11}[X_{10}] + (k_{-8} + k_8)[X_{45}],$

$a_{4,42} = k_{34}[X_6] + k_{-6}[X_{46}], \quad a_{4,44} = k_{31}[X_5],$

$a_{5,42} = k_5[X_6], \quad a_{5,44} = -k_{31}[X_5],$

$a_{64} = k_8[X_{45}] - k_6[X_6], \quad a_{65} = k_{31}[X_{44}],$

$a_{66} = -k_{-8}[X_4] - (k_5 + k_{-10}k_{19}[X_{18}])[X_{12}] - k_5([X_{42}] + [X_4]),$

$a_{6,42} = k_{-6}[X_{45}] - k_{34}[X_6],$

$a_{7,45} = k_{-8}[X_4] + k_3[X_{42}],$

$d_{42,4} = (k_6 + k_8)[X_{45}], \quad d_{42,5} = k_{31}[X_{44}],$

$d_{42,7} = k_{-6}, \quad d_{42,42} = (k_5 + k_{34})[X_6] + (k_{-6} + k_{34})[X_{46}],$

$d_{44,6} = k_5[X_{42}], \quad d_{44,44} = k_5[X_6] - k_{31}[X_5],$

$d_{45,6} = k_{34}[X_{42}], \quad d_{45,7} = k_8, \quad d_{45,45} = -(k_{-8} + k_8)[X_4],$

$d_{46,6} = k_6[X_4], \quad d_{46,7} = k_8, \quad d_{46,46} = -(k_{-6} + k_{34})[X_{42}].$

In case of uncompetitive positive allosteric modulation the vector of reacting components is $X_{Am}(t) = (X_1, ..., X_{34}, X_{42}, ..., X_{47})^T$, where $X_{47} := L_c \cdot R^T$ and the meaning of other components is as described above. The vector of constant concentrations of substrates remains unchanged. The matrix D_{Am} has new elements

$a_{44} = k_2([E_0] - [X_8]) + k_6[X_5] + (k_{-8} + k_5 + k_6)[X_6] + k_{-11}[X_{10}] - k_{-6}[X_{43}],$

$a_{4,42} = k_{-6}[X_{47}], \quad a_{4,44} = k_{-6}[X_8],$

$a_{54} = k_{-6}[X_{43}], \quad a_{55} = (k_{-5} + k_6)[X_4] - k_{33}[X_{42}],$

$a_{5,42} = k_{-6}[X_{47}], \quad a_{7,44} = k_{32}[X_8],$

$a_{88} = -k_{-6} - k_{-7} - (k_{-6} + k_{32})[X_{44}], \quad a_{8,42} = k_{33}[X_4], \quad a_{8,43} = k_6[X_{44}],$

$d_{42,42} = -k_{33}[X_5] + (k_{-6} + k_{33})[X_{47}], \quad d_{42,43} = k_{-6}[X_4], \quad d_{42,44} = k_{-6}[X_8],$

$d_{43,42} = k_3[X_5], \quad d_{44,43} = -(k_{-6} + k_6)[X_4], \quad d_{43,44} = k_{-6}[X_8],$

$d_{44,7} = k_{-7}, \quad d_{44,42} = k_{33}[X_{47}], \quad d_{44,44} = -(k_{-6} + k_{32})[X_8] + k_6[X_{43}].$

Initial concentrations of the reacting components close the system.

8.6   Model of PDE-4 Inhibitor

Cyclic nucleotide phosphodiesterases are the enzymes catalyzing the hydrolysis and inactivation of the second messengers, cyclic adenosine monophosphate, and cyclic guanosine monophosphate. Phosphodiesterase inhibitors potentially can increase signaling by inhibiting the cAMP enzyme breakdown. Nonspecific PDE inhibitors, such as papaverine, isobutylmethyl-xanthine, theophylline, etc., have been used by clinicians to treat cardiovascular problems, pulmonary hypertension and to facilitate some behavioral performance. Rolipram, a type 4 specific phosphodiesterase inhibitor, has been found to be effective in the management of patients with inflammation-induced preterm delivery. Results of in vitro experiments on chorionic cells demonstrated that the drug reduces the proinflammatory cytokine tumor necrosis factor alpha (TNFα) production and the activation of the transcription nuclear factor kappa B (NF-κB) (Hervé et al. 2008). Combination therapy of rolipram and a commonly used tocolytic terbutaline, β_2-adrenoceptor agonist, has shown a synergistic relaxing effect of drugs on the rat uterus in vitro (Klukovits et al. 2010). Experiments on human myometrial cells in culture confirmed the anti-inflammatory properties of PDE4 inhibitors in gestational tissues and their ability to prevent preterm delivery (Mehat et al. 2007).

Let L_{PDE4} be the specific reversible PDE-4 inhibitor. The proposed state diagram of its association with the enzyme is given in Fig. 8.4

The corresponding system of equations of chemical reactions is

$$dX_{PDE}/dt = D_{PDE}X_{PDE}(t) + C_{0,PDE}. \tag{8.9}$$

Here $X_{Am}(t) = \left(X_1, ..., X_{26}, X_{28}, X_{29}, ..., X_{34}, X_{48}\right)^T$ where new components have been introduced: $X_{28} := PDE, X_{48} := L_{PDE4} \cdot PDE, X_{49} := L_{PDE4},$, and the vector $C_{0,PDE} = (0, ..., C_{28}, C_{30}, 0, ...0, C_{34}, C_{48})^T$ where $C_{28} = C_{48} = k_{-35}[X_{49}]_0$.

The matrix D_{PDE} can be obtained from D where the following elements have been changed and added

$$b_{33} = -k_{23}[X_{48}],$$
$$b_{77} = -k_{35}[X_{48}] - k_{23}[X_{23}], \qquad b_{7,36} = -k_{-35},$$
$$d_{36,36} = -k_{-35} - k_{35}[X_{28}].$$

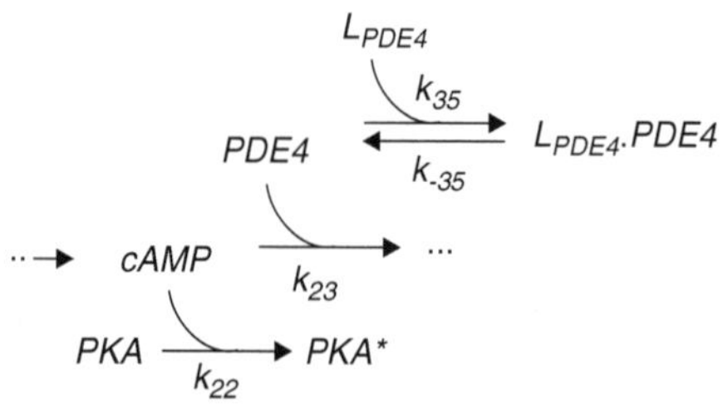

Fig. 8.4 The state diagram of signal transduction in myometrium in presence of a specific PDE-4 antagonist

In the above derivations we assumed that the drug is always available, i.e. $[L_{PDE4}] = [X_{49}]_0 = $ constant. Hence, the dynamics of its conversion in the system can be calculated from an algebraic equation

$$[X_{49}](t) = [X_{49}]_0 - [X_{48}](t). \tag{8.10}$$

Again, to close the system initial concentrations of the reacting substrates should be provided.

Exercises

1. Identify simulation strategies for drug discovery and drug development that are available today.
2. Signal transduction pathways are interconnected to form large networks. To guarantee signal specificity, i.e. intracellular signal to cell response, within such systems is difficult. What are the mechanisms that provide specificity of signaling?
3. Structure-guided drug design has become the preferred technique for the discovery of new pharmaceuticals. Study of binding of potential drugs to target receptors is an essential step in this process. This is often referred to as "the lock and key" model for protein–ligand docking. Although the latest technology has provided an unprecedented insight into the microscopic picture of the behavior of drug binding, why is there little progress in achieving desired biological activity?
4. Study the Monod, Wyman, and Changeux model of allosteric transition (Monod et al. 1965).
5. Current thinking in drug discovery suggests that tight ligand–receptor binding alone is not sufficient to separate the desired pharmacological effect from the mass-action noise of the biological system. What are the mechanisms involved?
6. A potential drawback of competitive inhibitors is that their efficiency and potency can be diminished by mass-action competition with the substrate. How can it be avoided?
7. Modify the mathematical model of PDE-4 inhibition (Fig. 8.4) that would include cyclic guanosine monophosphate as a second messenger system.
8. Advances in genomic research fuelled the expectation that highly specific drugs targeting a single disease-causing molecule could cure many complex diseases. Although some such drugs have proven to be successful, many have been found to be ineffective or to cause side effects. What are the limitations of the single-target drug paradigm?
9. The most successful drug targets are G-protein-coupled receptors, nuclear receptors, ion channels, and enzymes. Why has there been more success with these targets than others?
10. How can the pathway framework assist in drug discovery?

11. With the multiscale and large-volume data generation in the life sciences, drug discovery has become increasingly complicated. What is the role of a *virtual laboratory* environment in modern drug discovery and drug development? (Rauwerda et al. 2006)
12. Why are we unsuccessful in the design of drugs to manage myometrial dysfunctions?

Chapter 9
Gravid Uterus as a Soft Biological Shell

There are no such things as applied sciences, only applications of science.

L. Pasteur

9.1 Fundamental Assumptions

The one-dimensional model of the myometrial fasciculus studied in Chap. 3 provided an insight into the dynamics of electromechanical coupling among smooth muscle cells in the tissue. However, the results cannot be extrapolated to the whole organ to study the propagation of the wave of depolarization and associated contractions in the gravid uterus during labor and delivery. Here, we shall formulate a mathematical model of the uterus as a thin soft biological shell.

Based on experimental data and clinical observations obtained during and after delivery that:

1. Thickness of the wall of the pregnant organ varies from $h_1 = 0.4$–0.45 cm in the lower part of the body and from $h_2 = 0.5$–0.69 cm in the fundal region; the characteristic radii of the curvature of the middle surface in these regions range within $8 \leq R_1 \leq 9$ cm and $10 \leq R_2 \leq 13.8$ cm, respectively; thus max $(h_i/R_i) \leq 1/20$ $(i = 1, 2)$.
2. At full term, the height of the uterus measures on average 40 cm, $h/L = 10^{-2}$.
3. As the womb empties its fluid content, $\check{V} \simeq 600$ ml, at the end of the first stage of labor, intrauterine pressure (p) falls and the organ changes its shape and size as a result of rising tonus in the myometrium; however, in case of uterine atony – flaccid myometrium – the uterus "collapses" and can attain multiple configurations.
4. Stress–strain curves of uniaxial stretching in vitro experiments of myometrial strips, obtained from different parts of the nongravid and gravid uteri, have shown their high nonlinearity and large deformability, 40–100% (Pearsall and Roberts 1978) (at the time at which this book was written no experimental data

R.N. Miftahof and H.G. Nam, *Biomechanics of the Gravid Human Uterus*,   129
DOI 10.1007/978-3-642-21473-8_9, © Springer-Verlag Berlin Heidelberg 2011

on in-plane tensile mechanical properties of the myometrium under biaxial loading in vivo and in vitro were available).

It becomes clear that the pregnant uterus satisfies all the hypotheses and assumptions of the theory of thin soft shells (see Chap. 5). Furthermore, the wall of the organ, its geometry (configuration) and stress–strain distribution within it can be approximated, with a sufficient degree of accuracy, by the middle surface of the organ. Thus, henceforth it will be modeled as a thin soft biological shell (bioshell) (Fig. 9.1). The adjective "biological" is added to emphasize the nature of processes that are responsible for biomechanical events that take place.

Myometrial contractions in the gravid uterus are triggered by underlying electrical events in myocytes. The principle electrophysiological phenomena assume that:

5. Human myometrium is composed of two interspersed muscle strata – longitudinal and circumferential – with fasciculi running orthogonally within them.
6. The myometrium at term endows the properties of weak electrical anisotropy (longitudinal layer) and isotropy (circumferential layer), respectively (Wolfs and Rottinghuis 1970; Planes et al. 1984); multiple gap junctions, composed of connexin proteins of types 40, 43 and 45, provide the low electrical resistance gating among myocytes.
7. The electrical activity, either single or bursts of spikes, represents the integrated function of ion channels, presumably: voltage-dependent Ca^{2+} channels of L- and T-types, large Ca^{2+}-activated K^+ channels (BK_{Ca}), potential sensitive

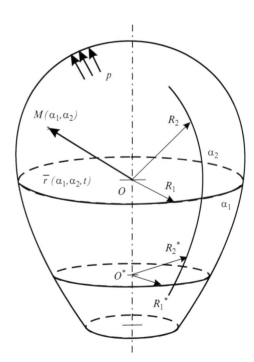

Fig. 9.1 The gravid human uterus as a thin bioshell

K$^+$ channels, and Cl$^-$ channels; the properties of the channels are neurohormonally modulated; this effect is assumed to be mainly chronotropic with an increase in the time of permeability for specific ions.

8. Although pacemaker cells per se have not been found in the organ, there are localized pacemaker zones near the uterotubal junctions that are responsible for generation of electrical signals and coordinated rhythmic contractions of labor.

9. Contractions are involuntary and, for the most part, independent of external control; internal regulation is provided by multiple neurotransmitters and hormones; all chemical reactions of substrate transformation satisfy first-order Michaelis–Menten kinetics.

Electrical events initiate a cascade of processes known as electromechanical coupling. The key player in it is free cytosolic calcium. Upon stimulation the concentration of Ca$_i^{2+}$ rises, mainly as a result of influx and partly following the mobilization of internal stores. We shall assume in calculations that:

10. The active force $T^a - $ Ca$_i^{2+}$ activity relationship is a generalized approximation of the experimental curves which are characteristic for smooth muscle is given by (3.6).

11. The total force $T_{c,l}$ generated by the myometrial layers is the result of deformation of its passive elements, $T^p(\lambda_c, \lambda_1, c_i)$, i.e. collagen and elastin fibers, and active contraction–relaxation of myofibrils, $T^a\left(\lambda_{c,l}, Z_{mn}^{(*)}, [\text{Ca}_i^{2+}], c_i\right)$, and is defined as

$$T_{c,l} = T^p(\lambda_c, \lambda_1, c_i) + T^a\left(\lambda_{c,l}, Z_{mn}^{(*)}, [\text{Ca}_i^{2+}], c_i\right), \tag{9.1}$$

where the meaning of parameters and constants are as described above.

Electrochemical and chemoelectrical coupling in the system is guaranteed by separate and/or conjoint activation of multiple intracellular pathways, and ligand-dependent transmembrane ion channels. The dynamics of their activation and inhibition is described by (7.18), (7.21) and (7.25), respectively.

The working assumptions (1)–(11) have been incorporated in a mathematical model of the pregnant uterus at term as discussed below.

9.2 Model of the Gravid Uterus

Let $\overset{0}{S}$ be the cut configuration of the bioshell that describes the shape of the gravid uterus with accuracy of bending in absence of loads. The first two-proliferative and synthetic phases in the organ are characterized by the extensive growth, remodeling, and homeomorphic change in shape without its actual deformation. Hence, it is reasonable to assume that during this stage the cut configuration coincides with the undeformed, S, configuration $(\overset{0}{S} = S)$ and:

(a) Stretch ratios $\lambda_{i,j} \equiv 1.0$, $(i,j = 1, 2)$;
(b) In-plane total forces $T_{ij}(\lambda_{i,j}) \equiv 0$ throughout the bioshell
(c) Intrauterine volume is $V \simeq 450 - 600\,\text{ml}$;
(d) Intrauterine pressure $p = 0$.

The pregnant uterus attains the deformed $\overset{*}{S}$ state with $\lambda_{i,j} > 1.0$, $T_{ij} > 0$, $\overset{\smile}{V} \simeq 800\,\text{ml}$ and $p = 2\text{--}12\,\text{mmHg}$ only toward the end of the second and, sustains it through the entire contractile phase.

Let the middle surface of the pregnant uterus be associated with a cylindrical coordinate system $\{r, \varphi, z\}$

$$
\begin{aligned}
r &= r(\alpha_1, \alpha_2, t), \\
\varphi &= \varphi(\alpha_1, \alpha_2, t), \\
z &= z(\alpha_1, \alpha_2, t),
\end{aligned}
\tag{9.2}
$$

that is related to the orthogonal Cartesian coordinates $\{x_1, x_2, x_3\}$ as

$$
x_1 = r \cos \varphi, \quad x_2 = r \sin \varphi, \quad x_3 = z.
$$

The position vector $\bar{r}$ of point $M(r, \varphi, z) \in S$ is given by

$$
\bar{r} = r\bar{k}_1 + z\bar{k}_3,
\tag{9.3}
$$

where

$$
\bar{k}_1 = \bar{i}_1 \cos \varphi + \bar{i}_2 \sin \varphi, \quad \bar{k}_2 = -\bar{i}_1 \sin \varphi + \bar{i}_2 \cos \varphi.
\tag{9.4}
$$

Differentiating (9.3) with respect to α_i with the help of (9.4), we obtain

$$
\bar{r}_i = \frac{\partial \bar{r}}{\partial \alpha_i} = \frac{\partial r}{\partial \alpha_i} \bar{k}_1 + r \frac{\partial \bar{k}_1}{\partial \alpha_i} + \frac{\partial z}{\partial \alpha_i} \bar{k}_3 = \frac{\partial r}{\partial \alpha_i} \bar{k}_1 + r \frac{\partial \varphi}{\partial \alpha_i} \bar{k}_2 + \frac{\partial z}{\partial \alpha_i} \bar{k}_3,
\tag{9.5}
$$

where

$$
\frac{\partial \bar{k}_1}{\partial \alpha_i} = \frac{\partial}{\partial \alpha_i} (\bar{i}_1 \cos \varphi + \bar{i}_2 \sin \varphi) = \frac{\partial \varphi}{\partial \alpha_i} (-\bar{i} \sin \varphi + \bar{i}_2 \cos \varphi) = \frac{\partial \varphi}{\partial \alpha_i} \bar{k}_2.
\tag{9.6}
$$

Projections of $\bar{r}_i$ in the direction of the r, φ, and z axes are given by

$$
\bar{r}_{ir} = \frac{\partial r}{\partial \alpha_i}, \quad \bar{r}_{i\varphi} = \frac{\partial \varphi}{\partial \alpha_i}, \quad \bar{r}_{iz} = \frac{\partial z}{\partial \alpha_i}.
$$

Hence,

$$
\bar{r}_i = \bar{r}_{ik} \bar{k}_1 + \bar{r}_{i\varphi} \bar{k}_2 + \bar{r}_{iz} \bar{k}_3.
\tag{9.7}
$$

Decomposing $\bar{e}_i$ along the base $\{\bar{k}_1, \bar{k}_2, \bar{k}_3\}$, we find

$$
\bar{e}_i = l_{ir} \bar{k}_1 + l_{i\varphi} \bar{k}_2 + l_{iz} \bar{k}_3,
\tag{9.8}
$$

here the direction cosines $l_{ij}(i = 1, 2, 3; j = r, \varphi, z)$ are given by

$$\bar{e}_1\bar{e}_2 = l_{1r}l_{2r} + l_{1\varphi}l_{2\varphi} + l_{1z}l_{2z} = \cos\overset{*}{\chi} = \overset{*}{a}_{12}/\sqrt{\overset{*}{a}_{11}\overset{*}{a}_{22}}$$
$$\bar{e}_i\bar{e}_3 = l_{ir}l_{3r} + l_{i\varphi}l_{3\varphi} + l_{iz}l_{3z} = 0, \qquad\qquad\qquad (9.9)$$
$$\bar{e}_n\bar{e}_n = l_{n1}^2 + l_{n2}^2 + l_{n3}^2 = 1, \qquad (n = 1, 2, 3)$$

Expanding the resultant of internal uterine pressure, $\bar{p}$, and the mass force, $\bar{f}$, in the direction of unit vectors $\bar{e}_i$, we get

$$\bar{p} = \bar{e}_1 p_1 + \bar{e}_2 p_2 + \bar{e}_3 p_3, \qquad\qquad (9.10)$$

$$\bar{f} = \bar{k}_1 f_r + \bar{k}_2 f_\varphi + \bar{k}_3 f_z. \qquad\qquad (9.11)$$

The vector of acceleration $\bar{a}(a_r, a_\varphi, a_z)$ in cylindrical coordinates is given by

$$a_r = -\frac{d^2 r}{dt^2} - r\left(\frac{d\varphi}{dt}\right)^2, \quad a_\varphi = -r\frac{d^2\varphi}{dt^2} + 2r\frac{dr}{dt}\frac{d\varphi}{dt}, \quad a_z = -\frac{d^2 z}{dt^2}. \quad (9.12)$$

Substituting (9.8), (9.10)–(9.12) in (5.63), the equations of motion of the soft shell in cylindrical coordinates take the form

$$\rho\sqrt{a}\left(\frac{d^2 r}{dt^2} - r\left(\frac{d\varphi}{dt}\right)^2\right) = \frac{\partial}{\partial\alpha_1}\left[(T^{1r}l_{1r} + T^{2r}l_{2r})\sqrt{\overset{*}{a}_{22}}\right]$$
$$+ \frac{\partial}{\partial\alpha_2}\left[(T^{1r}l_{1r} + T^{2r}l_{2r})\sqrt{\overset{*}{a}_{11}}\right] + (p_1 l_{1r} + p_2 l_{2r})\sqrt{\overset{*}{a}}$$
$$+ p_3(l_{2\varphi}l_{2z} - l_{1z}l_{2\varphi})\sqrt{\overset{*}{a}_{11}\overset{*}{a}_{22}} + \rho f_r\sqrt{a},$$

$$\rho\sqrt{a}\left(r\frac{d^2\varphi}{dt^2} + 2r\frac{dr}{dt}\frac{d\varphi}{dt}\right) = \frac{\partial}{\partial\alpha_1}\left[(T^{1\varphi}l_{1\varphi} + T^{1\varphi}l_{2\varphi})\sqrt{\overset{*}{a}_{22}}\right]$$
$$+ \frac{\partial}{\partial\alpha_2}\left[(T^{1\varphi}l_{1\varphi} + T^{2\varphi}l_{2\varphi})\sqrt{\overset{*}{a}_{11}}\right] + (p_1 l_{1\varphi} + p_2 l_{2\varphi})\sqrt{\overset{*}{a}}$$
$$+ p_3(l_{1z}l_{2r} - l_{1r}l_{2z})\sqrt{\overset{*}{a}_{11}\overset{*}{a}_{22}} + \rho f_\varphi\sqrt{a},$$

$$\rho\sqrt{a}\frac{d^2 z}{dt^2} = \frac{\partial}{\partial\alpha_1}\left[(T^{1z}l_{1z} + T^{2z}l_{2z})\sqrt{\overset{*}{a}_{22}}\right]$$
$$+ \frac{\partial}{\partial\alpha_2}\left[(T^{1z}l_{1z} + T^{2z}l_{2z})\sqrt{\overset{*}{a}_{11}}\right] + (p_1 l_{1z} + p_2 l_{2z})\sqrt{\overset{*}{a}}$$
$$+ p_3(l_{1r}l_{2\varphi} - l_{1\varphi}l_{2r})\sqrt{\overset{*}{a}_{11}\overset{*}{a}_{22}} + \rho f_z\sqrt{a}.$$
$$(9.13)$$

The dynamics of bursting and oscillatory myoelectrical activity, $V_{c,l}$, in the myometrial longitudinal (subscript l) and circumferential (c) layers, respectively, are adequately described by the system of equations (3.7)–(3.11).

The pacemaker activity in myometrial cells, V_p, is given by

$$\alpha C_m^p \frac{dV_p}{dt} = -\sum_j I_j + I_{ext(i)}, \qquad (9.14)$$

where C_m^p is the specific myometrial cell membrane capacitance, I_j are respective ion currents (j = L- and T-type Ca^{2+}, BK_{Ca}, voltage-dependent Na^+, K^+ and Cl^-), $I_{ext(i)} = V_i/R_p$ is the external membrane current, and R_p is the input cellular resistance. The currents are defined by

$$
\begin{aligned}
I_{Ca} &= \frac{g_{Ca(i)} z_{Ca}\left(V_p - V_{Ca}\right)}{1 + \vartheta_{Ca}\left[Ca_i^{2+}\right]}, \qquad
I_{Ca-K} = \frac{g_{Ca-K(i)}\rho_\infty\left(V_p - V_{Ca-K}\right)}{0.5 + \left[Ca_i^{2+}\right]}, \\
I_{Na} &= g_{Na(i)} m_{Na}^3 h_{Na}\left(V_p - V_{Na}\right), \qquad
I_K = g_{K(i)} n_K^4\left(V_p - V_K\right), \\
I_{Cl} &= g_{Cl(i)}\left(V_p - V_{Cl}\right),
\end{aligned}
\qquad (9.15)
$$

where $V_{Ca}, V_{Ca-K}, V_{Na}, V_K, V_{Cl}$ and $g_{Ca(i)}, g_{Ca-K(i)}, g_{Na(i)}, g_{K(i)}, g_{Cl(i)}$ are the reversal potentials are the maximal conductances of the respective ion currents, ϑ_{Ca} is the numerical parameter, and the turnover of free intracellular calcium yields

$$\frac{d\left[Ca_i^{2+}\right]}{dt} = \frac{0.2 z_{Ca}\left(V_p - V_{Ca}\right)}{1 + \vartheta_{Ca}\left[Ca_i^{2+}\right]} - 0.3\left[Ca_i^{2+}\right] \qquad (9.16)$$

The dynamic variables of the ion channels $z_{Ca}, \rho_\infty, m_{Na}, h_{Na}, n_K$ are given by

$$
\begin{aligned}
dz_{Ca}/dt &= (z_\infty - z_{Ca})/\tau_z, \\
dh_{Na}/dt &= \lambda_h(h_\infty - h_{Na})/\tau_h, \\
dn_K/dt &= \lambda_n(n_\infty - n_K)/\tau_n, \\
m_{Na} &= m_\infty(V_p), \quad \rho_\infty = \left[1 + \exp 0.15(V_p + 47)\right]^{-1}.
\end{aligned}
\qquad (9.17)
$$

In the above $m_\infty, h_\infty, n_\infty, z_\infty$ are calculated as

$$
\begin{aligned}
y_\infty &= \frac{\alpha_{y\infty}}{\alpha_{y\infty} + \beta_{y\infty}}, \quad (y = m, h, n) \\
z_\infty &= \left[1 + \exp(-0.15(V + 42))\right]^{-1},
\end{aligned}
\qquad (9.18)
$$

where

$$
\begin{aligned}
\alpha_{m\infty} &= \frac{0.12\left(V_p + 27\right)}{1 - \exp\left(-\left(V_p + 27\right)/8\right)}, \qquad
\beta_{m\infty} = 4\exp\left(-\left(V_p + 47\right)/15\right), \\
\alpha_{h\infty} &= 0.07\exp\left(-\left(V_p + 47\right)/17\right), \qquad
\beta_{h\infty} = \left[1 + \exp\left(-\left(V_p + 22\right)/8\right)\right]^{-1}, \\
\alpha_{n\infty} &= \frac{0.012\left(V_p + 12\right)}{1 - \exp\left(-\left(V_p + 12\right)/8\right)}, \qquad
\beta_{n\infty} = 0.125\exp\left(-\left(V_p + 20\right)/67\right).
\end{aligned}
$$

The propagation of the electrical wave of depolarization, $V_{c,l}^s$, along the myometrium in case of generalized electrical anisotropy is given by (6.33)

$$C_{\mathrm{m}}\frac{\partial V_{\mathrm{c,l}}^{s}}{\partial t} = I_{\mathrm{m1}}(\alpha_1, \alpha_2) + I_{\mathrm{m2}}(\alpha_1 - \alpha_1', \alpha_2 - \alpha_2') + I_{\mathrm{ion}}, \tag{9.19}$$

where $I_{\mathrm{m1}}, I_{\mathrm{m2}}$ are the transmembrane currents described by (6.29) and (6.32). Since the intra- and extracellular conductivity $\hat{g}_{\mathrm{i(o)}}$ of the syncytium is $\hat{g}_{\mathrm{i(o)}} := 1/R_{\mathrm{i(o)}}^{\mathrm{m}}$, where $R_{\mathrm{i(o)}}^{\mathrm{m}}$ is the intra-(subscript i) and extracellular (o) smooth muscle membrane resistance is $R_{\mathrm{i(o)}}^{\mathrm{m}} = R_{\mathrm{i(o)}}^{\mathrm{ms}}\lambda_{\mathrm{c,l}}/\tilde{S}_{\mathrm{c,l}}$, where $\lambda_{\mathrm{c,l}}$ are the stretch ratios and $\tilde{S}_{\mathrm{c,l}}$ are the cross-sectional areas of muscle layers, $R_{\mathrm{i(o)}}^{\mathrm{ms}}$ is the specific smooth muscle cell membrane resistance then assuming that $\tilde{S}_{\mathrm{c,l}}$ is constant throughout deformation of the myometrium, for $\hat{g}_{\mathrm{i(o)}}$ we have

$$\hat{g}_{\mathrm{i(o)}} = \frac{\tilde{S}_{\mathrm{c,l}}}{R_{\mathrm{i(o)}}^{\mathrm{m}}\lambda_{\mathrm{c,l}}} := \frac{\hat{g}_{\mathrm{i(o)}}^{*}}{\lambda_{\mathrm{c,l}}}. \tag{9.20}$$

Here $\hat{g}_{\mathrm{i(o)}}^{*}$ have the meaning of maximal intracellular and interstitial space conductivities. Substituting (9.20) into (6.29), (6.32), we obtain, finally,

$$I_{\mathrm{m1}}(\alpha_1, \alpha_2) = M_{vs}\left\{ \frac{2(\mu_{\alpha_2} - \mu_{\alpha_1})}{(1+\mu_{\alpha_1})(1+\mu_{\alpha_1})} \tan^{-1}\left(\frac{\mathrm{d}\alpha_1}{\mathrm{d}\alpha_2}\sqrt{\frac{G_{\alpha_2}}{G_{\alpha_1}}} \right) + \frac{\hat{g}_{0,\alpha_2}^{*}}{G_{\alpha_1}} \right\}$$
$$\times \left(\frac{\partial}{\partial \alpha_1}\left(\frac{\hat{g}_{0,\alpha_1}^{*}}{\lambda_c}\frac{\partial V_{\mathrm{l}}^{s}}{\partial \alpha_1} \right) + \frac{\partial}{\partial \alpha_2}\left(\frac{\hat{g}_{0,\alpha_2}^{*}}{\lambda_l}\frac{\partial V_{\mathrm{l}}^{s}}{\partial \alpha_2} \right) \right),$$

$$I_{\mathrm{m2}}(\alpha_1, \alpha_2) = M_{vs}\iint\limits_{S} \frac{(\mu_{\alpha_1} - \mu_{\alpha_2})}{2\pi(1+\mu_{\alpha_1})(1+\mu_{\alpha_1})}\frac{(\alpha_2 - \alpha_2')/G_{\tilde{s}_2} - (\alpha_1 - \alpha_1')/G_{\alpha_1}}{[(\alpha_1 - \alpha_1')/G_{\tilde{s}_1} - (\alpha_2 - \alpha_2')/G_{\alpha_2}]^2}$$
$$\times \left(\frac{\partial}{\partial \alpha_1}\left(\frac{\hat{g}_{0,\alpha_1}^{*}}{\lambda_c}\frac{\partial V_{\mathrm{l}}^{s}}{\partial \alpha_1} \right) + \frac{\partial}{\partial \alpha_2}\left(\frac{\hat{g}_{0,\alpha s2}^{*}}{\lambda_l}\frac{\partial V_{\mathrm{l}}^{s}}{\partial \alpha_2} \right) \right)\mathrm{d}\alpha_1'\mathrm{d}\alpha_2',$$

$$\mu_{\alpha_1} = \hat{g}_{0,\alpha_1}^{*}/\hat{g}_{\mathrm{i},\alpha_1}^{*}, \quad \mu_{\alpha_2} = \hat{g}_{0,\alpha_2}^{*}/\hat{g}_{\mathrm{i},\alpha_2}^{*},$$

$$G_{\alpha_1} = \frac{\hat{g}_{0,\alpha_1}^{*} + \hat{g}_{\mathrm{i},\alpha_1}^{*}}{\lambda_c}, G_{\alpha_2} = \frac{\hat{g}_{0,\alpha_2}^{*} + \hat{g}_{\mathrm{i},\alpha_2}^{*}}{\lambda_l}, G = \sqrt{G_{\alpha_1}G_{\alpha_2}}, \tag{9.21}$$

where α_1, α_2 are Lagrange coordinates of the longitudinal and circular smooth muscle syncytia, respectively, and the meaning of other parameters as described above.

The total ion current I_{ion} is given by

$$I_{\mathrm{ion}} = \tilde{g}_{\mathrm{Na}}\hat{m}^3\hat{h}(V_{\mathrm{c,l}}^{s} - \tilde{V}_{\mathrm{Na}}) + \tilde{g}_{\mathrm{K}}\hat{n}^4(V_{\mathrm{c,l}}^{s} - \tilde{V}_{\mathrm{K}}) + \tilde{g}_{\mathrm{Cl}}(V_{\mathrm{c,l}}^{s} - \tilde{V}_{\mathrm{Cl}}), \tag{9.22}$$

where $\tilde{g}_{\mathrm{Na}}, \tilde{g}_{\mathrm{K}}, \tilde{g}_{\mathrm{Cl}}$ represent maximal conductances, $\tilde{V}_{\mathrm{Na}}, \tilde{V}_{\mathrm{K}}, \tilde{V}_{\mathrm{Cl}}$ — reversal potentials of Na^+, K^+, and Cl^- currents. The dynamics of the probability variables, the activation and deactivation parameters of the ion gates are given by (3.14) and (3.15), respectively.

In case of myoelectrical isotropy of the syncytium, the governing equation obtains a simpler form

$$C_m \frac{\partial V_{c,1}^s}{\partial t} = \frac{M_{vs}}{1+\mu_{\alpha_1}} \left\{ \frac{\partial}{\partial \alpha_1} \left(\frac{\hat{g}_{0,\alpha_1}^*}{\lambda_c} \frac{\partial V_{c,1}^s}{\partial \alpha_1} \right) + \frac{\partial}{\partial \alpha_2} \left(\frac{\hat{g}_{0,\alpha_1}^*}{\lambda_1} \frac{\partial V_{c,1}^s}{\partial \alpha_2} \right) \right\} - I_{ion}. \quad (9.23)$$

Neurohormonal regulatory processes in myometrium are described by

$$d\mathbf{X}/dt = \mathbf{D}\mathbf{X}(t) + \mathbf{C_0}, \quad (9.24)$$

where the matrix $\mathbf{D}$ and vectors of reacting substrates $\mathbf{X}$ and $\mathbf{C_0}$ are given by (7.10), (7.18), (7.21) and the algebraic relationships (7.19), (7.22).

The following anatomically and physiologically justifiable initial and boundary conditions assume that:

12. The initial undeformed configuration of the pregnant uterus is defined by the intrauterine pressure, p, and the organ is at rest, i.e. myoelectrically quiescent.
13. The excitation of known intensity, V_p, and duration, t_i^d, is provided by electrical discharges in the pacemaker regions.
14. Concentrations of the reacting substrates are known.
15. The cervical end is electrically unexcitable and is either rigidly fixed or remains pliable throughout deformation; the condition of periodicity is imposed with respect to the angular coordinate, φ.

Thus, we have:

$$\text{at } t = 0 : V_p = \begin{cases} 0, 0 < t < t_i^d \\ V_p^0, t \geq t_i^d \end{cases}, \qquad V_{c,1} = V_{c,1}^r, \qquad V_{c,1}^s = 0,$$

$$[Ca_i^{2+}] = [Ca_i^{2+}]_0, \qquad \mathbf{X}(0) = \mathbf{X_0}, \qquad \mathbf{C}(0) = \mathbf{C_0},$$

$$r(\alpha_1, \alpha_2) = r_0(\alpha_1, \alpha_2), \quad \varphi(\alpha_1, \alpha_2) = \varphi_0(\alpha_1, \alpha_2), \quad z(\alpha_1, \alpha_2) = z_0(\alpha_1, \alpha_2),$$

and the dynamic variables of ion channels involved are defined by

$$\hat{m} = \hat{m}_\infty, \ \hat{h} = \hat{h}_\infty, \quad \hat{n} = \hat{n}_\infty, \ z_{Ca} = z_{Ca\infty}, \ h_{Na} = h_{Na\infty}, \ n_K = n_{K\infty},$$
$$\tilde{h} = \tilde{h}_\infty, \ \tilde{n} = \tilde{n}_\infty, \ \tilde{x}_{Ca} = \tilde{x}_{Ca\infty}, \quad (9.25)$$

for $t > 0$:

$$r(\alpha_1, \alpha_2)_{\alpha_1 = 0, L} = r_0(0, \alpha_2) = r_0(L, \alpha_2),$$
$$\varphi(\alpha_1, \alpha_2)_{\alpha_1 = 0, L} = \varphi_0(0, \alpha_2) = \varphi_0(L, \alpha_2),$$
$$z(\alpha_1, \alpha_2)_{\alpha_1 = 0, L} = z_0(0, \alpha_2) = z_0(L, \alpha_2), \quad (9.26)$$
$$V_{c,1}^s(0, \alpha_2) = V_{c,1}^s(L, \alpha_2).$$

To close the problem (9.13), (9.14), (9.19) and/or (9.23) should be complemented by constitutive relations (9.1). Finally, the mathematical model as formulated above describes:

(a) The pregnant uterus at term as a soft myoelectrically active biological shell
(b) Pacemaker activity in the organ
(c) The generation and propagation of electrical wave of excitation in the myometrial syncytia
(d) Electromechanical coupling and neurohormonal regulation of contractile activity.

9.3 Numerical Simulations

In the following paragraphs, results of numerical experiments of the study of the statics and dynamics of changes in shape, force–deformation distribution, electromechanical activity and pharmacological modulation of the gravid human uterus are presented and discussed. A diligent reader must have already recognized that the model contains numerous parameters and constants that have not yet been evaluated experimentally. For example, no information is available on mechanical constants c_i of the human uterus under biaxial loading, many constants of chemical reactions k_i also have so far not been estimated. To overcome the problem of missing input data, during simulations the myometrium is regarded as a curvilinear anisotropic nonlinear viscoelastic material with constitutive parameters derived from other soft human tissues. Thus, the uniaxial passive force–stretch ratio $T^p(\lambda, c_i)$ relationship lay bare "pseudoelastic" behavior; the active force function exhibits nonlinear dependence on the intracellular calcium concentration; and the biaxial in-plane $T^{ij}(\lambda_i, \lambda_j)$ function is a generalized form of the uniaxial constitutive relationship (Fig. 9.2). Therefore, the results of modeling are not aimed to achieve an accurate quantitative representation but instead offer a qualitative assessment of complex processes in the uterus during different stages of labor.

Labor is a series of continuous, progressive contractions of the uterus which help the cervix to open (dilate) and to thin (efface), allowing the fetus to move through the birth canal. They are divided into three stages: first stage begins with the onset of true labor and ends when the cervix is completely dilated. It starts with low intensity, short in duration (30–45 s) and irregular (5–20 min apart) contractions that progressively increase in strength, duration (60–90 s) and become regular and more frequent occurring every 0.5–1 min. At the end of first stage, the cervix dilates to 10 cm and the amniotic sac ruptures releasing $\tilde{V} \simeq 600$ ml of fluid. This leads to the fall in uterine pressure: $p = 2$–4 mmHg. The womb changes its shape and size as a result of rising tonus in myometrium.

The subsequent second stage is characterized by strong pushing contractions and expelling the baby through the birth canal to the outside world. The uterus contracts extensively during this stage and the intrauterine pressure rises to $p = 60$–100 mmHg. Shortly after the baby is born and the placenta is delivered – the

Fig. 9.2 General form
of uniaxial (**a**) and biaxial
(**b**) force–stretch ratio
relationships for soft
biological tissues
(myometrium)

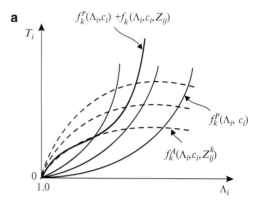

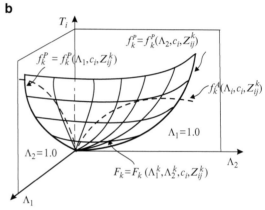

third final stage of labor – the uterus undergoes significant reduction in size. The
axial and transverse dimensions of the organ decrease by 20–30% and are attributed
to prevailing myometrial contractions. These mechanical and concomitant hormonal changes cause vasoconstriction and thus prevent postpartum bleeding.

9.3.1   Uterus Close to Term

The actual configuration and in-plane force–strain distribution in the pregnant
womb depends on: (1) the fetal presentation – vertex (cephalic), oblique or
transverse, (2) the fetal size, (3) the amount of amniotic fluid, and (4) the position
and size of the placenta. Although the current mathematical formulation does
not account explicitly each of these factors, their joint contribution in the model,
as a first approximation, can be attained through the variation of intrauterine
pressure p.

The configuration of the electrically quiescent human uterus at 38 weeks, which corresponds to end of the proliferative phase of gestation, represents the initial undeformed state of the bioshell. It is a result of the computer reconstruction of actual MRI data. Throughout simulations only a singleton pregnancy is considered.

Coronal view of the pear-like womb with the fetus in cephalic presentation at the beginning of the latent phase of first stage of labor is shown in Fig. 9.3. Analysis of the static total force distribution in the bioshell indicates that the fundus and the body of the organ experience excessive longitudinal total forces, while the lower segment predominantly circumferential. The cervix is closed with min $T_{c,l} \simeq 0$ produced in the region.

The period between 38 and 40 weeks is associated with low-intensity irregular contractions, known as Braxton–Hicks. An increase in myometrial tonus leads to smoothing and shortening of the lower segment. The shape of the uterus becomes more rounded with a decrease in the fundal height. The redistribution and decrease in the spatial gradient of T_l along with a significant rise in T_c in the lower segment are observed. There is a simultaneous increase in the in-plane forces T_c and T_l in the cervical region (Fig. 9.3).

In case of oblique presentation, the initial configuration of the uterus is skewed and elongated in the direction of the crown–rump axis of the fetus (Fig. 9.4). The body and the fundus undergo maximal tension in the longitudinal direction. The lower segment is unequally stretched with the presence of focal zones of excessive circumferential forces in that region. This pattern of $T_{c,l}$ distribution persists throughout the entire period of Braxton–Hicks contractions. Compared to cephalic presentation, the cervical region is subjected to intense T_c forces and $T_l \approx 0$. They increase gradually with gestation and attain the maximum at the end of 40th week.

Shapes and total force distributions in the uterus when the fetus is in transverse position are shown in Fig. 9.5. Results demonstrate that there is a constant increase in intensity of tension from 38 to 40 weeks in the longitudinal direction in the body and fundus of the organ. Small forces T_l are produced in the lower segment and the cervical region. The lower segment is overstressed circumferentially with max T_c exceeding similar values for cephalic and oblique presentations. The circumferential force in the area of the cervix remains low, $T_c \approx 0$, throughout the progression of pregnancy.

An increase in the intrauterine pressure above normal values, which occurs in polyhydramnious, and changes in mechanical characteristics of myometrium – ubiquitous stiffening or softening of it – do not affect the qualitative pattern of force–stretch ratio distribution in pregnant uteri.

A comparative analysis of forces at term in the cervical region depending on fetal presentation and the assumption that the myometrium is physically and statistically homogeneous are shown in Fig. 9.6. Results demonstrate that in the cephalic presentation, the cervix experiences pulling T_l and stretching T_c forces. Surprisingly, with the fetus in oblique presentation, the cervical region undergoes intense circumferential stretching and in transverse presentation – relatively small

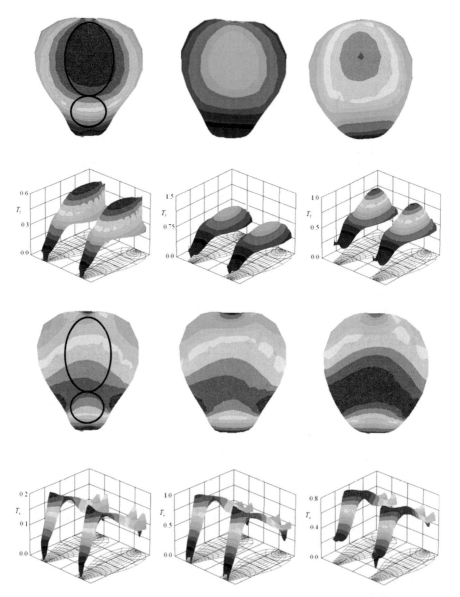

Fig. 9.3 Coronal views and corresponding total in-plane force distributions in the longitudinal, T_l, and circumferential, T_c, myometrial striata in the pregnant gravid human uterus at term with the fetus in cephalic position. Henceforth, all forces are normalized to the maximum force generated by the uterus through the whole case of study

and physiologically insignificant $T_{c,l}$ forces are being produced. Such conditions play definite roles in the dynamics of cervical opening during labor as discussed below.

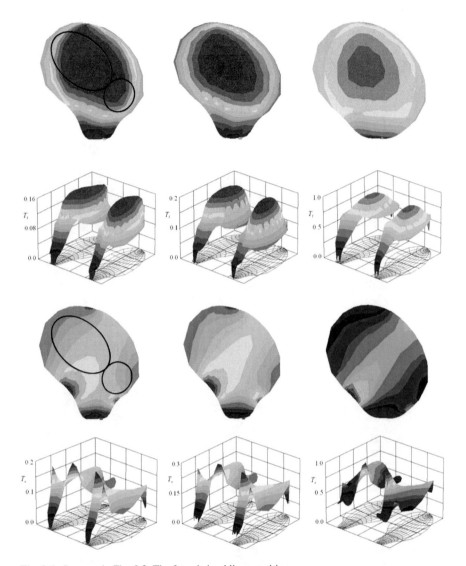

Fig. 9.4 Same as in Fig. 9.3. The fetus is in oblique position

9.3.2 First Stage of Labor

At the end of 40th week, the uterus enters first stage of labor which is marked
by irregular contractions. Let two separate pacemaker zones in the longitudinal
and circular layers are located in the uretotubal junctions. They discharge syn-
chronously multiple impulses of constant amplitude $\overset{0}{V_p} = 100\,\text{mV}$ and duration
$t_i^d = 10\,\text{ms}$ that generate electrical waves of depolarization V_1 and V_c, respectively.
The wave V_1 quickly spreads within the longitudinal fibers and encases a narrow

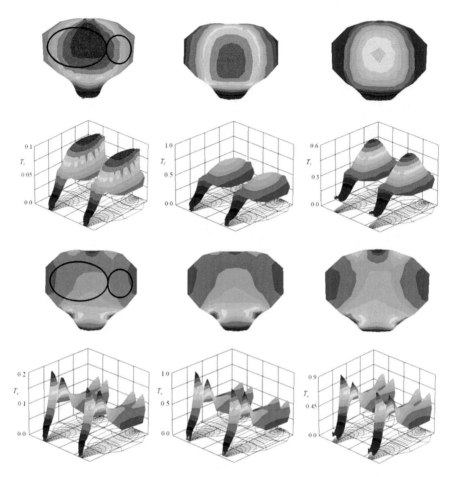

Fig. 9.5 Same as in Fig. 9.3. The fetus is in transverse position

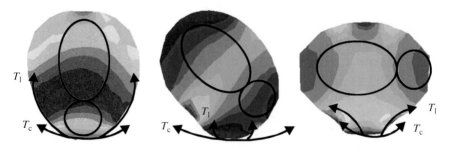

Fig. 9.6 The intensity of total stretch forces, T_1 and T_c, produced in the cervical region as a result of the fetal presentation

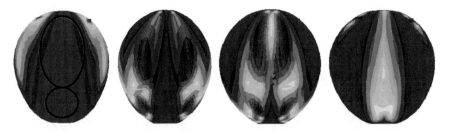

Fig. 9.7 Dynamics of the first stage of labor. Propagation of the electrical waves of depolarization $V_{l,c}$ within the longitudinal and circular muscle layers of the uterus

zone along the lateral sides of the womb (Fig. 9.7). As the two separate waves reach the lower segment, they begin to propagate circumferentially. The fronts of the waves of excitation collide in the region of the fundus and the body of the uterus with generation of a single solitary wave V_l.

The wave V_c within the circular syncytium extends circumferentially from the site of origin toward the lower segment. It provides a uniform excitation to the organ. Suffice it to note that the fundus and the body of the uterus experience strong depolarization compared to the lower segment and the cervical region.

The waves $V_{c,l}$ activate voltage-dependent calcium channels on the muscle membrane that results in a rapid influx of extracellular Ca^{2+} inside cells. A rise in the free cytosolic calcium ion concentration leads to activation of the cascade of mechanochemical reactions with production of the active forces of contraction (Fig. 9.8). Their pattern of propagation resembles the dynamics of spread of electrical waves. The most intense and lasting contractions are produced in the fundus and the body of the uterus. The lower segment is subjected to strong circumferential forces that assist the dilation of the cervix.

The uterus undergoes biaxial stretching throughout the entire first stage. Analysis of passive force distribution in the wall of the bioshell shows that the stroma experiences the most intense tension in both directions in the body of the organ. It is related to the changes in shape of the region in the anterior–posterior dimension from rounded to flatten (Fig. 9.9).

The pattern of total force distribution in the womb demonstrates the axial advancement, i.e. from the fundus to the body and the lower segment, and steady increase in the intensity of T_c. The dynamics is consistent with generation of the active forces of contraction by the myometrium (Fig. 9.10). Maximum T_c that evenly encircles the body and the lower segment of the uterus are observed close to the end of contraction. The force generated in the longitudinal striatum complements the process with the maximum T_l produced in the body along the anterior wall of the organ.

It is important to note that during the first stage of labor, the electromechanical wave processes sustain symmetry with respect to the axial dimension of the uterus and synchronicity of its pacemaker discharges. As a result, contractions cause the rupture of the amniotic sac and the release of 500–600 mL of fluid, the dilation of the cervix, the change in shape, and decrease in the fundal height of the organ.

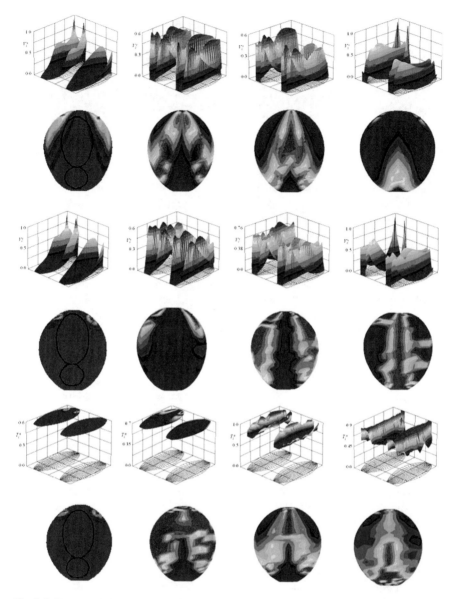

Fig. 9.8 Dynamics of the first stage of labor. Active force, $T^{\mathrm{a}}_{\mathrm{l,c}}$, distribution in the pregnant uterus

9.3.3   Second Stage of Labor

The subsequent second stage of labor is characterized by frequent strong uterine contractions aided by maternally controlled "pushing down" efforts. They typically

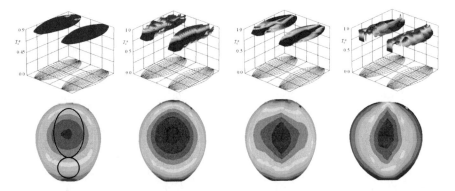

Fig. 9.9 Dynamics of the first stage of labor. Total force, $T_{1,c}$, distribution in the pregnant uterus

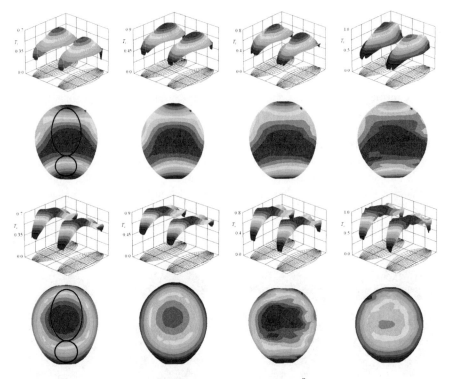

Fig. 9.10 Dynamics of the first stage of labor. Passive force, $T_{1,c}^{p}$, distribution in the pregnant uterus

occur before, during and after peak contraction force and are known as the "triple" and "peak" pushing styles. During this stage, the fetus is expelled from the womb. The process is associated with a series of internal deflections, translation and rotations of the head and the body of the fetus, and follows a specific trajectory – the curve of Carus.[1]

The succeeding results of simulation of the first stage of labor show that if the symmetry and synchronicity in electromechanical activity continue to prevail in the second stage, the fetus will undergo rapid translation only. No torque will be generated to initiate necessary rotations. Although currently there is no experimental evidence to support the thought, it is reasonable to believe that the pattern of electrical activity changes. Therefore, assume that the longitudinal syncytium is excited by synchronous discharges while there is a time-delay between dischargers of the left and right pacemaker zones and the stimulation of the circular syncytium. During numerical experiments, the intensity of impulses was increased twofold compared to the first stage of labor.

The shape of the uterus with a fully dilated cervix supported by internal pressure $p = 19$ mmHg is given. The propagation of the wave of depolarization V_1, passive T_1^p, active T_1^a and total force T_1 distribution in the organ are similar to those observed in the earlier stage and are not given here.

The discharge of the left pacemaker zone precedes the firing of the pacemaker on the right. The anterior front of the wave V_c originated at the left uterotubal junction envelops spirally the uterus before it collides with the similar wave that propagates from the opposite right uterotubal junction (Fig. 9.11).

The formed single wave continues to spread toward the lower segment and down to the cervical region.

The myometrium produces intensive contractions of duration ~1.5 min. They are concomitant with oscillations of intracellular calcium, $Ca_i^{2+}(t)$. As the pattern of active force T_c^a distribution corresponds to the dynamics of the wave of depolarization V_c, it is reasonable to assume the propagating mechanical wave of contraction produces a torque moment that rotates the fetus. Additionally, the total in-plane forces $T_{c,l}$ generated in the fundus and the body support its continual translation and later expulsion (Fig. 9.12).

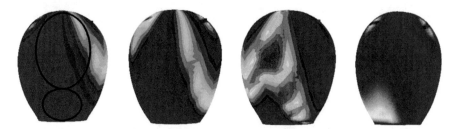

Fig. 9.11 Dynamics of the second stage of labor. Propagation of the electrical waves of depolarization V_c within the circular myometrium

---

[1]A twisted and curved line representing the outlet of the pelvic canal. The end of the curve is at the right angle to its beginning.

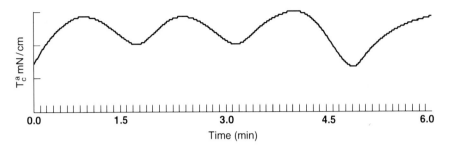

Fig. 9.12 Typical trace of active force T_c^a progression in the circular myometrium layer of the pregnant uterus

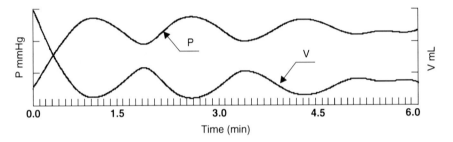

Fig. 9.13 Changes in the intrauterine pressure and volume during the second stage of normal labor (the fetus is in cephalic presentation)

Changes in volume of the uterus with every contraction reflect changes in intrauterine pressure (Fig. 9.13). It increases up to five times from the resting value and peaks with the maximum of $T_{c,l}^a$ and $T_{c,l}$. Interestingly, the rate $dp/dt = const$ indicates that the myometrium at this stage behaves as pure elastic biomaterial.

9.3.4 Third Stage of Labor

The third stage of labor is characterized by a few regular strong contractions that separate the placenta from the wall and then deliver it through the birth canal. There is an associated increase in myometrial tonus, thickening of the uterine wall (average $h \geq 2.4$ cm) (Herman et al. 1993; Deyer et al. 2000), significant reduction in size – the fundal height measures ~15–20 cm, and flattening of the organ in the anterior–posterior dimension, ~8–12 cm. Detailed sonographic evaluation of in vivo changes in myometrial thickness after delivery showed that the fundus measures 27.37 ± 3.5 mm, the anterior wall -40.94 ± 3.5 mm, and the posterior wall -42.34 ± 2.44 mm, respectively (Buhimschi et al. 2003). The characteristic radii of the curvature of the middle surface of the uterus range within $15 \leq R_{1,2} \leq 20$ cm and $\max(h_i/R_i) >> 1/20$. Thus, the organ at this stage does not satisfy the hypotheses of thin shells and should be treated as a thick shell of variable thickness.

Although the extension of the current thin shell approach to model the postpartum uterus is beyond the scope of this book, we shall only highlight a number of distinct features of thick shells that significantly influence the appropriateness of the model and results of stress–strain distribution in the organ. First, the Kirchhoff–Love assumptions, i.e. plane sections remain plane after the deformation and perpendicular to the middle surface of the shell is no longer valid. As a result, the transverse shear strains cannot be neglected and the angle of rotation of the cross-section is altered. Second, the initial curvatures of the thick shell contribute to the generated stress resultants and stress couples, and cause a nonlinear distribution of the in-plane stresses across the thickness of the shell. This is because the length of the surface away from the middle surface is different from that of the middle surface. Third, in case of the uterus, it is important to consider the radial stress distribution over the thickness of the shell that is imperative from a physiological point of view.

At the moment when the book was written, there were no publications in scientific literature that addressed the above questions with regard to the postpartum uterus.

9.3.5 Constriction Ring

Constriction ring is a pathological condition that can occur at any stage of labor and is characterized by a persistent localized annular spasm of the circular myometrium. The exact pathophysiology of the phenomenon is not known. The predisposing factors though have been identified and include: malpositioning or malpresentation of the fetus and improper stimulation of the uterus with oxytocin. The constriction can develop at any part of the uterus but more frequently it appears at the junction of the body and the lower segment. This is in concert with the oxytocin hypothesis and the pattern of OT-receptor distribution in the womb. Critical analysis of the literature shows that during labor they are overexpressed primarily in the fundus and the body of the organ making them more susceptible to the hormone.

Consider the constriction ring – the constant active force T_c^a applied circumferentially – during the second stage of labor as shown in Fig. 9.14. The presence of a "muscular band" does not affect the dynamics of spread of the waves of depolarization V_c and V_l. They freely pass the constricted zone and reach the cervical region. The intensity of T_c^a in the organ varies with each contraction, while T_c^a remains constant at the site of the ring. The region above the constriction experiences contractions, however, the cervix remains lax: $T_{c,l}^a \approx 0$ and $T_{c,l} \approx 0$. The passive force – the reaction of the fibrilar stroma to stretch – at the zone of constriction is less compared to the values T_c^p observed during normal labor.

Addition of Atosiban™ to the system resulted in relaxation of the constriction ring and loss of productive contractions in the entire uterus.

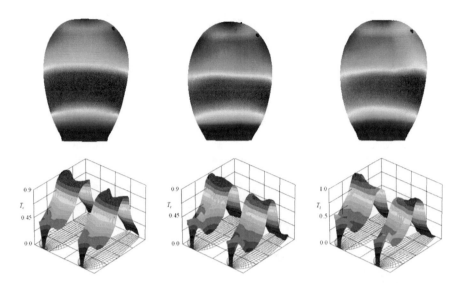

Fig. 9.14 Dynamics of total force, T_c, development in the circular myometrium in case of constriction ring

Although the fetus per se is not included in the model, we can speculate objectively about implications the constriction ring has on the dynamics of labor and delivery. Thus, ineffective mechanical activity and small forces generated during contractions may be associated with uterine dystocia, the delay in fetal descent, if the ring occurs at the level of the internal cervix or around the fetal neck, retention of the placenta, and postpartum hemorrhage.

9.3.6 Uterine Dystocia

Uterine dystocia (dysfunctional labor) refers to a difficult or prolonged labor and can be either cervical, i.e. the cervix does not efface and dilate, or uterine, e.g. constriction ring. Multiple causes including the abnormal fetal size and position, fetopelvic disproportion, mineral and electrolyte imbalance, uncoordinated electromechanical myometrial activity, abnormal rigidity of the cervix and cervical conglutination have been implicated in the pathogenesis of the condition.

Consider the gravid uterus in the first stage of labor. Let numerous, in addition to the ubiquitous two uterotubal, pacemaker zones be present and scattered over the organ. They fire impulses of constant amplitude $V_p^0 = 100\,\mathrm{mV}$ and duration $t_i^d = 10\,\mathrm{ms}$ at random. Assume also that the tissue of the cervix is rigid and nonstretchable.

Under the above settings, the induced waves of depolarization $V_{c,l}$ fail to propagate effectively along the longitudinal and circular myometrial syncytia

(Fig. 9.15). Signals propagate over short distances only. Multiple interferences and confluences prevent powerful excitation of the organ.

The amplitude of the active and the total in-plane forces generated by the uterus is low. Contractions of the circular and longitudinal muscle layers are not synchronized. The anterior front of the electromechanical wave becomes irregular and there is a loss of the directionality of its propagation – from the fundus toward the lower segment. As a result, the cervix fails to efface and dilate, and remains closed throughout. This significantly affects the dynamics of expulsion of the fetus which is notably impaired.

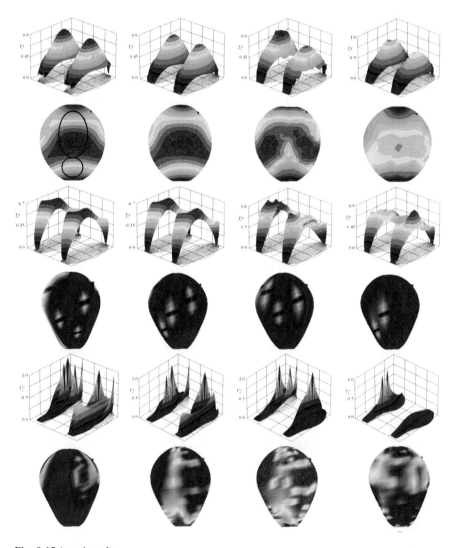

Fig. 9.15 (continued)

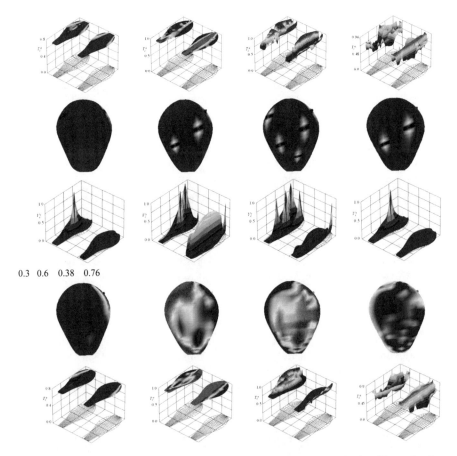

0.3 0.6   0.38   0.76

Fig. 9.15 Uterine dystocia. Propagation of the electrical waves of depolarization, $V_{l,c}$, and active force, $T^a_{l,c}$, development within the longitudinal and circular muscle layers of the uterus

9.3.7 Hyper- and Hypotonic Uterine Inertia

An elevated tone of the uterus that generally occurs in the latent phase of the first stage of labor is known as hypertonic uterine dysfunction. The uterus produces irregular, frequent, and strong contractions which are not effective. The etiological factors concerned are not very clear but the following conditions including the mechanically overreacting body of the womb and/or a lack of coordinated spread of the wave of depolarization and nerve pulse synchronization in the myometrium have been incriminated. Resting intrauterine pressure between contractions is high (>12 mmHg). The cervix dilates slowly and as a result the labor is prolonged.

To simulate the hypertonic uterine inertia assume there is a time lag between discharges of the uterotubal pacemaker zones that trigger electrical waves of excitation in the longitudinal and circular myometrium. Let also the frequency and amplitudes of impulses that are being produced by pacemakers exceed the normal values (Sect. 9.3.2).

Results of calculations clearly demonstrate that in the beginning the longitudinal myometrium of the fundus, the body, and the lower segment undergo uniform rigorous contractions. Their magnitude is higher compared to the values recorded during normal labor. In contrast, the circular myometrium generates intensive active forces mainly in the body and the lower segment while the strength and contractility of the fundal region remain low.

As the wave of depolarization quickly engulfs the uterus and the active and passive forces fully develop, the organ becomes almost evenly stressed – the total in-plane forces $T_{c,l}$ attain relatively constant values over the surface of the bioshell (Fig. 9.16). This effect is achieved by heterogeneous distribution of the forces $T_{c,l}^p$ and $T_{c,l}^a$ over different regions of the uterus. It is noteworthy that the cervix fails to dilate and stays contracted throughout.

Pethidine, a drug with analgesic and antispasmodic properties, is being used to treat hypertonic uterine inertia in clinics. It exerts its pharmacological action through κ-opioid and μACh receptors, and interactions with voltage-gated transmembrane Na^+ channels. Additionally, pethidine inhibits the dopamine and noradrenaline transporters and the uptake mechanisms.

The effect of the drug in the model was achieved by simulating conjointly the following pharmacokinetic mechanisms: muscarinic antagonism, adrenergic agonism, and decrease in the permeability of g_{Na}. Addition of pethidine into the system caused hyperpolarization and relaxation of myometrium.

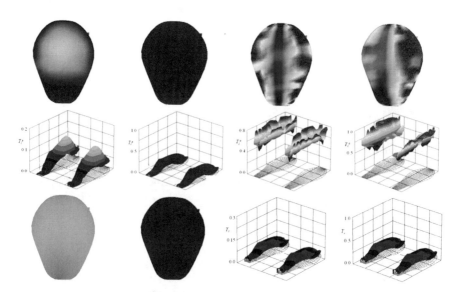

Fig. 9.16 Hypertonic uterine inertia. Dynamics of passive, $T_{l,c}^p$, active, $T_{l,c}^a$, and total force, $T_{l,c}$, development in the longitudinal and circular myometria

The infrequent, small amplitude, and of short duration contractions during labor indicate on hypotonic uterine inertia. Although the exact etiology of the uterine inertia remains undetermined, the condition appears to be related to a combination of factors. Among them are a deficiency of endogenous oxytocin and prostaglandins, an overdistension of the uterus, acquired or inherited pathology of myometrium, uterine hypolasia, and others. Two types of hypotonic inertia, i.e. primary and secondary, are recognized. Weak and low frequency contractions from the start of labor are characteristic of primary inertia while the substantial lessening of contractility after a period of good uterine electromechanical activity is suggestive of secondary hypotonic inertia.

A decrease in firing strength and rate of pacemaker zones leads to the loss of synchronization in depolarization of myometrium and, as a result, development of forces of active forces. The intensity of total forces $T_{c,l}$ is lower compared to normal labor. The shape of the organ stays unchanged and the cervix closed with each contraction.

Application of oxytocin has a distinct excitatory effect. The resting tonus of myometrium increases and strong forceful contractions of normal amplitude and high frequency are produced. Their dynamics is similar to the one observed during normal labor (Figs. 9.7–9.12) and are not given here.

Exercises

1. There is an ongoing debate about the organization of fasciculi in myometrium. One group proposes the double layer, while the opponents argue in favor of homogenous structure of the tissue. How would the assumption of homogeneity affect the governing system of equations?
2. Discuss the concept of weakly connected oscillators (Hoppensteadt and Izhikevich 1997) with regard to electrical activity of myometrium.
3. Use ABS Technologies® to study the effect of electrical isotropy of both the longitudinal and circular muscle layers on stress–strain distribution in the human uterus during labor.
4. Is the pregnant human uterus an autonomous system?
5. Verify (9.13).
6. Use ABS Technologies® to study the effect of mechanical heterogeneity of myometrium on dynamics of stress–strain development in the human uterus during labor.
7. Discuss clinical significance of the results presented in Fig. 9.6.
8. Use ABS Technologies® to study the effect of high-frequency discharges of pacemaker zones at uterotubal junctions on the dynamics of first stage of labor.
9. How might volitional pushing efforts affect the stress–strain distribution in the uterus during the second stage of labor?
10. Use ABS Technologies® to study the role of electrolyte imbalance in pathogenesis of uterine dystocia.
11. Successful remedies elicit their beneficial effects by targeting simultaneously and interacting with different receptors in regulatory extra- and intracellular pathways in a delicate way. Multicomponent therapies are increasingly gaining

attention in management of various diseases. Discuss the possible mechanisms that underlie the combinatorial effects of multicomponent therapies.

12. Use ABS Technologies® to suggest rational treatment of hypotonic uterine inertia, i.e. identify potential targets and study the effects of their modulations on uterine activity.

13. Discuss the general concept of multistage therapy with regard to the management of labor abnormalities (Kitano 2007).

Chapter 10
Biomechanics of the Gravid Uterus in Perspective

> In mathematics the art of proposing a question must be held of higher value than solving it.
>
> G. Cantor

10.1 Ontology of Models

Mathematics is a powerful tool to explore, understand, and predict reality. The mode of operation of mathematics is a model. It constitutes a theoretical projection based on comprehension of the event, accuracy of application of physical laws and principles, precision of mathematical formulation, and algorithms employed. In addition to summarizing a given set of hypothetical concepts, it also summons them into the mind of the user and demands that he/she conceptualizes a relation between the object of study and the model. This can bring the system in line with the concepts, or it can bring the concepts in line with the system.

The general trend in modern biology is toward quantitative and comprehensive investigation of biological systems. The explosion in experimental data at the molecular, sub-cellular and cellular levels that are difficult to interpret in relation to the physiological behavior of tissues and organs has demanded the development of new effective tools to integrate intricate physiological processes at multiscale levels. Conventional approaches have shown their ineptness in unraveling the elaborate pathways of hierarchical interactions, which is why we still have a limited perception of pathological mechanisms and, as a result, poor prediction and management of many diseases. Mathematical biology, on the other hand, a scientific discipline that combines empirical, mathematical and computational techniques, provides a coherent framework for analysis of the functionality of biological systems in terms of their morphostructural elements.

Systems biology modeling is referred to the application of systems theory to study the complexity of biological functions at various spatiotemporal scales by using a common representation. The complexity is related to a combination of different factors: (1) the limited knowledge of the underlying mechanisms, (2) the

R.N. Miftahof and H.G. Nam, *Biomechanics of the Gravid Human Uterus*,
DOI 10.1007/978-3-642-21473-8_10, © Springer-Verlag Berlin Heidelberg 2011

diversity of spatial and temporal scales, (3) the range of physicochemical reactions, (4) the high level of interdependence among functions, and (5) the intrinsic dynamic nonlinearity of processes. For example, the spatial scales span from the gene (~10^{-10} m) to the whole body (~1 m) and the temporal scales – from the dynamics of ion channels (~10^{-6} s) to a whole life (~10^{9} s); the level of interdependence emerge from gene and protein interactions with intra- and extracellular structures; nonlinear physicochemical phenomena arise from signal transduction cascades, metabolic pathways, coupled electromechanical processes, etc. Modeling and analysis of such systems is a formidable task that can only be accomplished effectively with significant computer (mathematical) input.

In the last few decades, modeling has been trying to explain high-level physiological functions at the level of genes. One of the major challenges for this approach is to identify the sites at which different functions are expressed. Although it is increasingly apparent that predictions to physiological outcomes depend on genomic knowledge, the gene-oriented approach has its dark side. Since multiple genes contribute to any given function, the number of all possible combinations is enormous ~$2e^{166713}$ (Crampin et al. 2001)! With such a large number of permutations, to understand the system purely in terms of its genome, i.e. genes and their interactions, is clearly impossible.

Proteomics and metabolomics is the level of ontology which encompass dynamic interactions among proteins in response to internal and external perturbations. It quantifies cellular functions in terms of translational and posttranscriptional modification rates and protein decay. The relation between the gene-oriented and protein-based approach is clearly seen in modeling of mRNAs expression profiles that encode fibronectin and α5-integrin receptor. These models offer a valuable insight to understand myometrial hypertrophy and remodeling during late pregnancy and to elucidate the force transduction during contractions. The role of nardilysin convertase 1 in regulation of a 44-amino acid (mK44) insertion in the structure of BK_{Ca} channels and their significance in controlling the excitability of myometrium have been emphasized in previous chapters (Korovkina et al. 2009). Another example concerns modeling of signaling pathways that mediate adaptive responses of the myometrial cytoskeleton to external perturbations. Responses are closely integrated with gene expression, focal adhesion proteins, c-Src, paxillin and small heat shock proteins, such as αB-crystallin and HSP27, that are found in membrane dense plaques (Gerthoffer and Gunst 2001; MacIntyre et al. 2008). Such associations will allow quantitative description and prediction of contractility of the uterus.

The in vivo identification and measurement of gene–protein and protein–protein interactions poses a great experimental problem. Proteins are less amenable to high-throughput studies than are nucleic acids and existing methods, e.g. two-dimensional poly acrylamide gel electrophoresis, micro-array and mass spectrometry, do not have sufficient accuracy and selectivity. Consequently, the scarcity of reliable data presents the main hurdle in the development of robust mathematical models at this level.

The next spatiotemporal scale is cell models. Until recently, mathematical models of a uterine myocytes have been concerned mainly with electrical processes, myofilament mechanics, and excitation–contraction coupling. Major advances in physiology of the autonomous nervous system in recent years have made it necessary to revise our thinking about the mechanisms of chemical signal transduction and modulation. A concept of colocalization and cotransmission by multiple transmitters, including monoamines, amino acids, peptides and hormones, has shifted the earlier dominating emphasis on "linear" perception of a stimulus – a single effect relationship toward a "nonlinear" view, e.g. a stimulus – various effects. Although it has had a profound influence on the understanding of physiology, of the pregnant human uterus few of the current models take these views into account.

There is a big knowledge gap between intercellular signaling and the role of receptor polymodality in the integration of biological responses. Constantly, evolving associations of receptors with intracellular pathways have shown to be crucial to the coordination of contractions. However, there is no experimental data on receptor distribution and receptor expression in the human uterus. Critical reevaluation of the literature revealed: strong dependence of regional electromechanical responses of the ewe uterus to α-adrenergic stimulation by phenylephrine on the time of gestation, i.e. up to 48 h before parturition, only the cervical end of the uterine horn reacted to the drug, while during labor the entire organ responded with contractions (Prud'Home 1986); the dominant role of β_3, and not β_2 and β_1, adrenoceptors, as was previously thought, in control of the relaxation of human myometrium (Rouget et al. 2005); increased expression of OT and $PGF_{2\alpha}$ receptors and responsiveness to OT and prostaglandin, respectively, in the fundus, and a selective increase in expression of OT receptors in the body but not in the lower segment of the gravid human uterus, which is more reactive to $PGF_{2\alpha}$ (Brodt-Eppley and Myatt 1999). The role of other transmitters (tachykinins, purines, serotonin, histamine) and receptors in pregnancy and labor is yet to be determined.

Another mode of cellular communication is electrical signaling. The basis for that is the conjoint activity of different transmembrane ion channels. Fundamental electrophysiological properties of the ion channels of myometrium have been identified and verified experimentally in vitro or by using the "knock-out" methodology. Results provide invaluable quantitative data on channel kinetics and show their role in sustaining quiescence and excitability of the gravid uterus. However, to extrapolate these findings to actively contracting cells at term and in labor requires a combined analysis of function of all channels simultaneously in relation to the environment they operate. For example, conductivity of BK_{Ca} channels should be evaluated in conjunction with deformation of caveolae; convincing experimental evidence suggests that stretching of the uterus triggers electrical response, however, little is known about the existence and function of stretch receptors and mechanosensitive ion channels; intricacies of ionotropic and metabotropic transduction mechanisms and the origin for pacemaker activity in the pregnant uterus have not yet been fully investigated. To understand the complexity and to predict the dynamics of relations among different sub-cellular and cellular components, models based on anatomically detailed quantitative information should be developed. Once built, they would be the key to diagnosing and managing uterine pathologies, which to this point have proved elusive.

Many physiological cell models contain considerable biophysical details. However, to link them with clinically relevant whole organ pathology requires integration at the tissue level. This could be done by invoking a continuum hypothesis which involves averaging over a sufficient number of cells present. It is important to note that the fine details are not lost in the process and can be reconstructed from the coarse scale solution. The advantage of such an approach is that it allows analyzing nano- and macroscale events within the unified theoretical framework of phenomenological entities (stress, strain, chemical concentrations, electrical potentials), and formulating constitutive relations in terms of functions of genes, proteins, and cells. It is noteworthy that constitutive relations describe the behavior of a tissue under conditions of interest, and not the tissue itself. For example, although a tissue may be best classified as mixture-composite that exhibits inelastic characteristics under particular environments, it may be modeled as elastic or viscoelastic body. Various models have been proposed to study cytoskeletal interconnectivity including percolation, tensegrity, soft glassy rheological, and continuum models. None of them has been widely accepted though to describe human myometrium. Given the diversity of histomorphological elements and the environment they function, one should expect that a combined description will be most successful. Hence, a model of extracellular matrix–integrin–cytoskeletal axis based on discrete molecular dynamics and a continuum description may help reveal how external mechanical forces are transduced into biochemical signals and induce biological responses.

The complexity of computational models increases tremendously as one moves to the higher organ level. To reproduce relevant behavior of an organ demands a continuous balanced iteration, i.e. inclusion vs. elimination, reduction vs. integration, selection vs. rejection, between experimental datasets and theoretical concepts. There are examples of promising attempts in developing biologically realistic models of the human heart (Noble 2002), the lung (Crampin et al. 2001), the kidney (Thomas 2009), and the digestive tract (Miftahof et al. 2009; Miftahof and Nam 2010). It is a great challenge in the coming years to construct a unified biologically plausible mathematical model of the pregnant human uterus and pelvic floor structures that would incorporate genes, proteins (contractile, anchoring, neurotransmitters, hormones), different types of cells and tissues (connective, muscle, vascular, nervous).

10.2   Applications, Pitfalls, and Problems

Obstetrics has become a highly quantitative, computer-intensive science. With the computing power to harness the data available, and with insights offered from mathematics, researchers have made fascinating progress in: interpreting noninvasive transabdominal electromyographic recordings obtained from pregnant women and correlating the electrical myometrial activity with contractions (Horoba et al. 2001; Garfield et al. 2005; Ramon et al. 2005a, b; Marque et al. 2007; La Rosa et al. 2008); analyzing the pregnant uterus under extreme impact loadings

(Jelen et al. 1998; Duma et al. 2005; Delotte et al. 2006; Acar and van Lopik 2009); simulating childbirth (Buttin et al. 2009; Lien et al. 2009; Li et al. 2009a, b, c; Jing et al. 2009); unraveling the function and role of the pelvic floor muscles during labor and delivery (d'Aulignac et al. 2005; Lee et al. 2009a, b; Lien et al. 2009; Li et al. 2009a, b, c; Li et al. 2010; Martins et al. 2007; Noakes et al. 2008; Weiss and Gardiner 2001; Calvo et al. 2009). Indeed, such applications are continuously expanding.

The modern misconception is to treat mathematics as if it had no limits or as if it never makes mistakes. Because mathematics is a product of mental exercise, it self-contains human errors, and along with the successful uses of it, there are many examples of abuses. Thus, the finite element method, a numerical technique, and Matlab®, a high-level language and interactive computational environment commonly substitute models of the uterus and surrounding anatomical structures. The misconception is brought about by the misuse of the essential characteristic, i.e. mesh discretization of a continuous domain into a set of sub-domains – elements – which is very attractive to a naïve modeler when dealing with noncanonical shapes like the human womb, baby's head, birth canal, and pelvic floor components. Mathematical formulations and governing equations in many studies are missing. Often investigators rely on a priori installed general models and mathematical formulations borrowed from the mechanics of solids, fluids, electromagnetism, etc., which do not reflect exactly processes they try to mirror.

Large-scale projects like childbirth call for a combined modeling of the uterus and pelvic floor structures. The anatomical and physiological complexity of the latter poses another twist of challenges to a mathematical modeler. MRI and CT imaging technologies provide invaluable detailed information about geometry and morphology of the organ (Hoyte and Damaser 2010). However, they do not supply quantitative data about their actual mechanical characteristics, i.e. deformability, elasticity, fatigability, muscle tone, etc. Models of vaginal childbirth have been proposed where the pelvic muscles were assumed to possess passive isotropic/ transversely isotropic, hyperelastic or viscoelastic characteristics analogous to exponential, Money-Rivlin or neo-Hookean materials. Attempts have been made to include active contractions by increasing the values of forces acting against the descending fetal head without significant improvement in the outcome results. While geometry of the fetal head was reproduced with high degree of anatomical accuracy, from a mechanical perspective, it was treated as a nondeformable solid body. This is a strict nonphysiological constraint because the head experiences large displacements while passing through the birth canal owing to mobility of parietal bones, in particular. It would be more appropriate to consider maternal–fetal interaction as a contact dynamic rather than a kinematic problem where translational and rotational displacements of the fetus and expulsion pressure profiles are prescribed. Although such models are impressively descriptive and attractive in presentation, they have a limited clinical application. To develop a feasible model of childbirth that describes the dynamics, provides an insight to intricate physiological mechanisms, foresees and advises on pharmacological corrections and interventions, if and when needed, is a great challenge and the ultimate goal of computational biology.

We have demonstrated that equations of the theory of thin shells could accurately describe the human uterus at term and during the first two stages of labor. However, the basic constructive assumptions become invalid during the third stage, when significant changes in the thickness of the organ with respect to other dimensions take place. Another concern is that myometrium is commonly treated as a physically homogeneous continuum. Morphological findings strongly indicate though that the fundus and the body are formed mainly of the smooth muscle, while the cervix of the connective tissue. Dynamic anatomical and functional links among cells and tissues guarantee the tensegrity which is paramount for normal function. Therefore, it would be more adequate to model the uterus as a laminated, regionally heterogeneous, nonlinear soft shell with the degree of structural anisotropy defined by the orientation of fasciculi. At the moment of writing the book, no systematic experimental studies have been conducted to investigate generalized active and passive mechanical properties of the human uterus and its adaptive changes – remodeling and reconstruction. Capture these structure–function relationships is the key to successful computational biomechanics of the organ.

Myometrium responds with contractions to electrical excitation. This guarantees electromechanical coupling – a valuable link between the structure and function in a way that no engineering material can emulate. Although the dynamics of the spread of excitation can be well captured using the Hodgkin–Huxley formalism, the origin and generation of pacemaker potentials in the gravid uterus remains a question of acute scientific debate. A solution to the problem will advance the understanding of the physiology of onset and the control of labor.

Models of the pregnant uterus and childbirth, without exception, lack the neuroendocrine signaling control system. Morphostructural elements of the uterus – cells – form large networks of signaling pathways. Despite extensive morphological and functional connections, the organ operates with high degree specificity and accuracy and any disturbance, either morphological or chemical, may lead to a disease state. The challenge is to unravel the "signaling specificity and stability" in the uterus during different stages of pregnancy and labor. This will increase clinical applicability of models, and perhaps most significantly, will help find new therapeutic modalities to treat uterine dysfunction and premature labor, in particular, which is the central problem in modern obstetrics.

If used appropriately, models can yield valuable insight and provide inaccessible information about the system under investigation. However, before a model can be applied to study a particular system, it has to be evaluated for accuracy of working hypotheses and robustness of the algorithm. This is achieved through a series of test simulations over a wide range of empirical conditions, with a subsequent qualitative and quantitative comparison of theoretical and experimental results. The process is highly dependent on the quality of input data, which has always been a problem with any biological model. As we have mentioned earlier, the lack of robust data about physiological, mechanical, chemical, etc., numerical parameters of the human pregnant uterus precludes the factual simulation of normal and pathological labor.

Mathematical formulations of biological models are usually presented as non-linear systems of partial and/or ordinary differential equations. In most cases, they can only be solved numerically. Often investigators use commercial software and packages as mathematical solvers that are easily available on the market and are being developed and released at a fast rate every day. Because they are written in different programming languages, for different computer architectures, and are designed for various applications, it is important to know how to make use of them sparingly. The choice should be based on clear comprehension of the mathematics of the problem and the proposed algorithm. The investigator may control the former but not the latter, which is customarily hidden and protected by the vendor. One must realize though that the numerical output has an aura of intellectual superiority and is capable to reassure beyond its real limits because "even wrong numbers may pose as true and convey an illusion of certainty and security that is not warranted" (Kitching et al. 2006).

The interpretation and extrapolation of theoretical results to the real system should be attempted only after the confidence in a model has been firmly established. Several models might be consistent with the data at hand and even yield the same mathematical representation and results. Consequently, the inadvertent pitfall emerges when the extrapolation of the similar to desirable and intuitively guessed solutions is attempted. A model is only good as its underlying assumptions and, hence, most models have a limited range of applicability. The developer is acutely aware of these limitations. But once a model is released in the public domain, it may be taken by others and used outside its range of validity. There are no definite recipes on how to prevent or avoid such mistakes. Each model should be transparent with regard to biological and working hypotheses, its mathematical description and the method of solution. Until all requirements are met and related problems are solved, any model of the pregnant uterus and childbirth will bear a sense of ambiguity and will be treated with a certain degree of doubt.

Finally, without the help of mathematical modeling and computer simulation our understanding of physiology of the pregnant human uterus during labor will "remain a prisoner of our inadequate and conflicting physical intuitions and metaphors" (Harris 1994). In the book, we have only touched on the challenges involved in developing and integrating models of the uterus. We hope that it serves to motivate and encourage interested analysts undertake new research into this fascinating and rapidly expanding field of computational biology.

An optimist may see a light where there is none, but why must the pessimist always run to blow it out?

(R. Descartes)

Exercises

1. The tools required for integrative modeling in biology are different from the ones used in standard engineering. Biological materials have a unique characteristic to grow and remodel in response to changing environment. This guarantees an important link between the structure and function in a way that

no engineering material can emulate. What are the "environmental" changes that determine uterine remodeling and growth?

2. The development of appropriate theoretical frameworks in growth and remodeling mechanics is crucial in modern biomechanics. The microstructural models for tissue consider the individual behaviors of the primary structural constituents (collagen and elastin fibers, muscle cells) and their spatial orientation. Discuss phenomenological assumptions for a microstructural model of myometrium as a heterogeneous electromechanical syncytium.

3. Constructing a model is something of an art. In the book, we have been concerned mainly with electromechanical activity and neurohormonal regulation of the pregnant human uterus. We have established the structure–function axis from the molecular to organ level. Consider the uterus with a fetus. Discuss ontology levels of the corresponding model.

4. All experiments present a compromise between relevance, reproducibility, ethical issues, and cost. Modern research in order to be successful requires experimentation at a multitude of levels of complexity. What are the roles, advantages, and disadvantages of theoretical (mathematical) models in experimental biology and medicine?

5. Doctors will always remain skeptical with regard to the application of computer models in clinical settings. One of the major reasons is that any model lacks patient-specificity. How can this problem be solved?

6. A great deal of attention is currently focused in reverse engineering dynamic models of regulatory networks. Scientists expect that in the future it may be possible to manipulate the genome in situ, and by selectively turning genes on and off in vivo using genetically engineered switches. What ethical issues might this approach present?

7. A framework for the myometrial physiome project has been put in place which involves developing ontologies for describing the biological knowledgebase, markup languages for encapsulating models of structure and function at all spatial scales. What are the challenges and factors that drive this project?

8. The simpler the mathematical model, the greater the understanding. If one were to incorporate all the currently available data on the human uterus into a mathematical model then it would be computationally impossible and would fail to provide the needed insight. How complex should the model be?

9. Aristotle (384–322 BC) suggested that "Here and elsewhere we shall not obtain the best insights into things until we actually see them growing from the beginning." In recent years, research efforts have focused on "reductionist" biology, i.e. unraveling mysteries and mechanisms of function at the level of genes and proteins, which has proven to be successful. Have we achieved the goal?

10. Should we even try to model the pregnant human uterus?

References

Acar BS, van Lopik D (2009) Computational pregnant occupant model, 'Expecting', for crash simulations. Proc Inst Mech Eng Part D J Automob Eng 223:891–902

Adaikan PG, Adebiyi A (2005) Effect of functional modulation of Ca^{2+}- activated Cl^- currents on gravid rat myometrial activity. Indian J Pharmacol 37:21–25

Aelen P (2005) Determination of the uterine pressure with electrodes on the abodomen. PhD Thesis, Eindhoven Univ Technol, p. 30

Åkerud A (2009) Uterine remodeling during pregnancy. PhD Thesis, Lund University, Sweden

Ambrus G, Rao CV (1994) Novel regulation of pregnant human myometrial smooth muscle cell gap junctions by human chorionic gonadotrophin. Endocrinology 135:2772–2779

Anwer K, Sanborn B (1989) Changes in intracellular free calcium in isolated myometrial cells: role of extracellular and intracellular calcium and possible involvement of guanine nucleotide-sensitive proteins. Endocrinology 124:17–23

Asboth G, Phaneuf S, Europe-Finner GN, Toth M, Lopez-Bernal A (1996) Prostaglandin E_2 activates phospholipase C and elevates intracellular calcium in cultured myometrial cells: involvement of EP1 and EP3 receptors subtypes. Endocrinology 137:2572–2579

Asboth G, Phaneuf S, Lopez-Bernal AL (1997) Prostaglandin receptors in myometrial cells. Acta Physiol Hung 85:39–50

Awad SS, Lamb HK, Morgan JM, Dunlop W, Gillespie JI (2007) Differential expression of ryanodine receptor RyR2 mRNA in the non-pregnant and pregnant human myometrium. Biochem J 322:777–783

Bardou M, Rouget C, Breuiller-Fouché M, Loustalot C, Naline E, Sagot P, Freedman R, Morcillo E, Advenier C, Leroy MJ, Morrison JJ (2007) Is the beta3-adrenoceptor (ADRB3) a potential target for uterorelaxant drugs? BMC Pregnancy Childbirth. doi:10.1186/1471-2393-7-S1-S14

Belmonte A, Ticconi C, Dolci S, Giorgi M, Zicari A, Lenzi A, Jannini EA, Piccione E (2005) Regulation of phosphodiesterase 5 expression and activity in human pregnant and non-pregnant myometrial cells by human chorionic gonadotrophin. J Soc Gynecol Investig 12:570–577

Benson AP, Clayton RH, Holden AV, Kharche S, Tong WC (2006) Endogenous driving and synchronization in cardiac and uterine virtual tissues: bifurcations and local coupling. Philos Transact A Math Phys Eng Sci 364:1313–1327

Bernal AL, TambyRaja RL (2000) Preterm labour. Best Pract Res Clin Obstet Gynaecol 14: 133–153

Blanks AM, Shmygol A, Thornton S (2007) Myometrial function in prematurity. Best Pract Res Clin Obstet Gynaecol 21:807–819

Brainard AM, Miller AJ, Martens JR, England S (2005) Maxi-K^+ channel localize to caveolae in human myometrium: a role for an actin-channel-caveolin complex in the regulation of myometrial smooth muscle K^+ current. Am J Physiol Cell Physiol 289:C49–C57

Brodt-Eppley J, Myatt L (1999) Prostaglandin receptors in lower segment myometrium during gestation and labor. Obstet Gynecol 93:89–93

Buhimchi CS, Buhimchi IA, Zhao G, Funai E, Peltecu G, Saade GR, Weiner CP (2007) Biomechanical properties of the lower uterine segment above and below the reflection of the urinary bladder flap. Obstet Gynecol 109(3):691–700

Buhimschi CS, Buhimschi IA, Malinow AM, Weiner CP (2003) Myometrial thickness during human labor and immediately post partum. Am J Obstet Gynecol 188(2):553–559

Burghardt RC, Barhoumi R, Sanborn BM, Andersen J (1999) Oxytocin-induced Ca^{2+} responses in human myometrial cells. Biol Reprod 60:777–782

Burnstock G (2006) Term-dependency of P2 receptor-mediated contractile responses of isolated human pregnant uterus. Eur J Obstet Gynecol 129:128–134

Burnstock J (2004) Cotransmission. Curr Opin Pharmacol 4:47–52

Burridge K, Chrzanowska-Wodnicka M (1996) Focal adhesions, contractility, and signaling. Ann Rev Cell Dev Biol 12:463–518

Bursztyn L, Eytan O, Jaffa AJ, Elad D (2007) Mathematical model of excitation-contraction in a uterine smooth muscle cell. Am J Physiol Cell Physiol 292:C1816–C1829

Buttin R, Zara F, Shariat B, Redarce T (2009) A biomechanical model of the female reproductive system and the fetus for the realization of a childbirth virtual simulator. Conf Proc IEEE Eng Med Biol Soc 2009:5263–5266

Cali JJ, Zwaagstra JC, Mons N, Cooper DMF, Krupinski J (1994) Type VIII adenylyl cyclase. A Ca^{2+}/calmodulin-stimulated enzyme expressed in discrete regions of rat brain. J Biol Chem 269:12190–12195

Calvo B, Pena E, Martins P, Mascarenhas T, Doblare M, Natal Jorge RM, Ferreira A (2009) On modelling damage process in vaginal tissue. J Biomech 42:642–652

Caulfield MP, Birdsall NJM (1998) Classification of muscarinic acetylcholine receptors. Pharmacol Rev 50:279–290

Celeste P, Mercer B (2008) Myometrial thickness according to uterine site, gestational age and prior cesarean delivery. J Matern Fetal Neonatal Med 21(4):247–250

Cha S (1968) A simple method for derivation of rate equations for enzyme-catalyzed reactions under the rapid equilibrium assumption or combined assumptions of equilibrium and steady state. J Biol Chem 25:820–825

Chan EC, Fraser S, Yin S, Yeo G, Kwek K, Fairclough RJ, Smith R (2002) Human myometrial genes are differentially expressed in labor: a suppression subtractive hybridization study. J Clin Endocrinol Metab 87:2435–2441

Chanrachakul B, Pipkin FB, Khasn RN (2004) Contribution of coupling between human myometrial beta2-adrenoreceptor and the BK_{Ca} channel to uterine quiescence. Am J Physiol Cell Physiol 287:C1747–C1752

Chapman NR, Kennelly MM, Harper KA, Europe-Finner GN, Robson SC (2006) Examining the spatio-temporal expression of mRNA encoding the membrane-bound progesterone receptor-alpha isoform in human cervix and myometrium during pregnancy and labour. Mol Hum Reprod 12:19–24

Chien EK, Saunders T, Phillippe M (1996) The mechanism underlying Bay K 8644-stimulated phasic myometrial contractions. J Soc Gynecol Investig 3:106–112

Claudine Serradeil-Le Gal C, Valette G, Foulon L, Germain G, Advenier C, Naline E, Bardou M, Martinolle J-P, Pouzet B, Raufaste D, Garcia C, Double-Cazanave E, Pauly M, Pascal M, Barbier A, Scatton B, Maffrand J-P, Le Fur G (2004) SSR126768A (4-Chloro-3-[(3R)-(+)-5-chloro-1-(2,4-dimethoxyben-zyl)-3-methyl-2-oxo-2,3-dihydro-1H-indol-3-yl]-N-ethyl-N-(3-pyridylmethyl)-benzamide, hydrochloride): a new selective and orally active oxytocin receptor antagonist for the prevention of preterm labor. J Pharmacol Exp Ther 309:414–424

Coleman RA, Smith WL, Narumiya S (1994) Internatinal Union of Pharmacology classification of prostanoid receptors:properties, distribution, and structure of the receptors and their subtypes. Pharmacol Rev 46:205–229

Collins PL, Moore JJ, Idriss E, Kulp TM (1996) Human fetal membranes inhibit calcium L-channel activated uterine contractions. Am J Obstet Gynecol 175:1173–1179

Conrad JT, Johnson WL, Kuhn WK, Hunter CA (1966a) Passive stretch relationships in human uterine muscle. Am J Obstet Gynecol 96:1055–1059

Conrad JT, Kuhn WK, Johnson WL (1966b) Stress relaxation in human uterine muscle. Am J Obstet Gynecol 95:254–265

Cordeaux Y, Pasupathy D, Bacon J, Charnock-Jones DS, Smith GCS (2009) Characterization of serotonin receptors in pregnant human myometrium. J Pharmacol Exp Ther 328:682–691

Crampin EJ, Haltstead M, Hunter P, Nielsen P, Noble D, Smith N, Tawhai M (2001) Computational physiology and the physiome project. Exp Physiol 89(1):1–256

Dahle LO, Andersson RG, Berg G, Hurtig M, Ryden G (1993) Alpha adrenergic receptors in human myometrium: changes during pregnancy. J Soc Gynecol Investig 36:75–80

d'Aulignac D, Martins JAC, Pires EB, Mascarenhas T, Jorge RMN (2005) A shell finnite element model of the pelvic floor muscles. Comput Methods Biomech Biomed Engin 8:339–347

Degani S, Leibovitz Z, Shapiro I, Gonen R, Ohel G (1998) Myometrial thickness in pregnancy: longitudinal sonographic study. J Ultrasound Med 10:661–665

Delotte J, Behr M, Baque P, Bourgeon A, Peretti F, Brunet C (2006) Modeling the pregnant woman in driving position. Surg Radiol Anat. doi:10.1007/s00276-006-0102-3

Deyer TW, Ashton-Miller JA, Van Baren PM, Pearlman MD (2000) Myometrial contractile strain at uteroplacental separation during parturition. Am J Obstet Gynecol 183(1):156–159

Doheny HC, Houlihan DD, Ravikumar N, Smith TJ, Morrison JJ (2003) Human chorionic gonadotrophin relaxation of human pregnant myometrium and activation of the BK_{Ca} channel. J Clin Endocrinol Metab 88:4310–4315

Duma S, Moorcroft D, Stitzel J, Duma G (2005) Computational modeling of a pregnant occupant. http://onlinepubs.trb.org/onlinepubs/conf/CP35v2.pdf

Elliott CL, Slatter DM, Denness W, Poston L, Bennett PR (2000) Interleukin 8 expression in human myometrium: changes in relation to labor onset and with gestational age. Am J Reprod Immunol 43:272–277

Engstrøm T, Bratholm P, Vihardt H, Christensen NJ (1999) Effect of oxytocin receptor and β_2-adrenoceptor blockade on myometrial oxytocin receptors in parturient rats. Biol Reprod 60:322–329

Eta A, Ambrus G, Rao CV (1994) Direct regulation of human myometrial contractions by human chorionic gonadotrophin. J Clin Endocrinol Metab 79:1582–1586

Eude-Le Parco IE, Dallot E, Breuiller-Fouche M (2007) BMC Pregnancy Childbirth. doi:10.1186/1471-2393-7S1-S11

Ferre F, Uzan M, Janssens Y, Tanguy G, Jolivet A, Breuiller M, Sureau C, Cedard L (1984) Oral administration of micronized natural progesterone in late human pregnancy. Effects on progesterone and estrogen concentrations in the plasma, placenta, and myometrium. Am J Obstet Gynecol 148:26–34

Fitzgibbon J, Morrison JJ, Smith TJ, O'Brien M (2009) Modulation of human smooth muscle cell collagen contractility by thrombin, Y-27632, TNF alpha and indomethacin. Reprod Biol Endocrinol. doi:10.1186/1477-7827-7-2

FitzHugh RA (1961) Impulses and physiological states in theoretical models of nerve membrane. Biophys J 79:917–1017

Flügger W, Chou SC (1967) Large deformation theory of shells of revolution. J Appl Mech 34:56–58

Fuchs AR, Fuchs F, Husslein P, Soloff MS (1984) Oxytocin receptors in the human uterus during pregnancy and parturition. Am J Obstet Gynecol 150(6):734–741

Gabella G (1984) Structural apparatus for force transmission in smooth muscle. Physiol Rev 64:455–477

Galimov KZ (1975) Foundations of the nonlinear theory of thin shells. Kazan State University, Kazan, Russia

Gao B, Gilman AG (1991) Cloning and expression of a widely distributed (type IV) adenylyl cyclase. Proc Natl Acad Sci USA 88:10178–10182

Garfield RE, Maner WL, Mackay LB, Schlembach D, Saade GR (2005) Comparing uterine electromyography activity of antepartum patients versus term labor patients. Am J Obstet Gynecol 193:23–29

Garfiled RE, Maner WL (2007) Physiology of electrical activity of uterine contractions. Semin Cell Dev Biol 18(3):289–295

Garfiled RE, Maner WL, MacKay LB, Schlembach D, Saade GR (2005) Comparing uterine electromyography activity of antepartum patients versus term labor patients. Am J Obstet Gynecol 193:23–29

Gerthoffer WT, Gunst S (2001) Focal adhesion and small heat shock proteins in the regulation of actin remodeling and contractility in smooth muscle. J Appl Physiol 91:963–972

Giannopoulos G, Jackson K, Kredester J, Tulchinski D (1985) Prostaglandin E and F_{2alpha} receptor in human myometrium during the menstrual cycle and in pregnancy and labor. Am J Obstet Gynecol 153:904–912

Gimpl G, Fahrenholz F (2001) The oxytocin receptor system: structure, function, and regulation. Physiol Rev 81:629–683

Goerttler K (1968) Structure of the human uterus wall. Arch Gynecol 205:334–342

Goupil E, Tassy D, Bourguet C, Quiniou C, Wisehart V, Pétrin D, Le Gouill C, Devost D, Zingg HH, Bouvier M, Saragovi HU, Chemtob S, Lubell WD, Claing A, Hébert TE, Laporte SA (2010) A novel biased allosteric compound inhibitor of parturition selectively impedes the prostaglandin F2α-mediated Rho/ROCK signaling pathway. J Biol Chem 285:25624–25636

Gunja-Smith Z, Woessner JF (1985) Content of the collagen and elastin cross-links pyridinoline and the desmonies in the human uterus in various reproductive states. Am J Obstet Gynecol 153: 92–95

Hai CM, Murphy RA (1988) Cross-bridge phosphorylation and regulation of latch state in smooth muscle. Am J Physiol Cell Physiol 254:C99–C106

Hamada Y, Nakaya Y, Hamada S, Kamada M, Aono T (1994) Activation of K channels by ritodrine hydrochloride in uterine smooth muscle cells from pregnant women. Eur J Pharmacol 288:45–51

Hanoune J, Pouline Y, Tzavara E, Shen T, Lipskaya L, Miyamoto N, Suzuki Y, Defer N (1997) Adeny6lyl cyclases: structure, regulation and function in an enzyme superfamily. Mol Biol Cell 8:2365–2378

Harris AK (1994) Multicellular mechanics in the creation of anatomical structures. In: Akkas N (ed) Biomechanics of active movement and division of cells. Springer, Berlin, pp 87–129

Hartzell C, Putzier I, Arreola J (2005) Calcium-activated chloride channels. Annu Rev Physiol 67:719–758

Heitman LH, Mulder-Krieger T, Spanjersberg RF, von Frijtag Drabbe Künzel JK, Dalpiaz A, IJzerman AP (2006) Allosteric modulation, thermodynamics and binding to wild-type and mutant (T277A) adenosine A_1 receptors of LUF5831, a novel nonadenosine-like agonist. Br J Pharmacol 147(5):533–541

Herman A, Weinraub Z, Bukovsky I, Arieli S, Zabow P, Caspi E, Ron-El R (1993) Dynamic ultrasonographic imaging of the third stage of labor: new perspectives into third-stage mechanisms. Am J Obstet Gynecol 168(5):1496–1499

Hertelendy F, Zakar T (2004) Regulation of myometrial smooth muscle functions. Curr Pharm Des 10:2499–2517

Hervé R, Schmitz T, Evain-Brion D, Cabrol D, Leroy M-J, Méhats C (2008) The PDE4 inhibitor rolipram prevents nuclear factor κB binding activity and proinflammatory cytokine release in human chorionic cells. J Immunol 181:2196–2202

Hoppenstedt FC, Izhikevich EM (1997) Weakly connected neural networks. Springer, New York

Horoba K, Jezewski J, Wrobel J, Graczyk S (2001) Algorithm for detection of uterine contractions from electrohysterogram. In: Proceedings, 23rd annual conference, IEEE/EMBS, October 25–28, Istanbul, Turkey

Hoyte L, Damaser MS (2010) Magnetic resonance-based female pelvic anatomy as relevant for maternal childbirth injury simulations. Obstet Gynecol 115:804–808

Ilic D, Damsky CH, Yamamoto T (1997) Focal adhesion kinase: at the crossroads of signal transduction. J Cell Sci 110:401–407

Inoue Y, Nakao K, Okabe K, Izumi H, Kanda S, Kitamura K, Kuriyama H (1990) Some electrical properties of human pregnant myometrium. Am J Obstet Gynecol 162:1090–1098

İrfanoğlu B, Karaesmen E (1993) A biomechanical model for the gravid uterus. In: Brebbia CA et al (eds) Trans Biomed Health, pp 59–65

Jacquemet V (2006) Pacemaker activity resulting from the coupling with unexcitable cells. Phys Rev E 74:011908

Jan LY, Jan YN (1994) Potassium channels and their evolving gates. Nature 371:119–122

Jelen K, Otahal S, Doleal A (1998) Response of a pregnant uterus to impact loading. In: Proceedings XVI International symposium on biomechanics in sports, pp 313–316

Jing D, Ashton-Miller J, DeLancey JO (2009) How different maternal volitional pushing affect the duration of the second stage of labor: a 3-D viscohyperelastic finite element model. http://me.engin.umich.edu/brl/

Joyner RW, Wilders R, Wagner MB (2006) Propagation of pacemaker activity. Med Biol Eng Comput. doi:10.10007/s11517-006-0102-9

Kao CY, McCullough JR (1975) Ion current sin the uterine smooth muscle. J Physiol 246(1):1–36

Karash YM (1970) Radiotelemetric investigation of fluctuations in intrauterine pressure during intervals between labor pains. Bull Exp Biol Med 70:861–863

Karteris E, Zervou S, Pang Y, Dong J, Hillhouse EW, Randeva HS, Thomas P (2006) Progesterone signaling in human myometrium through two novel membrane G protein coupled receptors: potential role in functional progesterone withdrawal. Mol Endocrinol 20:1519–1534

Kenakin T (2004) Principles: receptor theory in pharmacology. Trends Pharmacol Sci 25:186–192

Khac LD, Arnaudeau S, Lepretre N, Mironneau J, Harbon S (1996) Beta adrenergic receptor activation attenuates the generation of inositol phosphates in the pregnant rat myometrium. Correlation with inhibition of Ca^{2+} influx, a cAMP-independent mechanism. J Pharmacol Exp Ther 276:130–136

Khan RN, Matharoo-Ball B, Arulkuraman S, Ashfold MLJ (2001) Potassium channels in the human myometrium. Exp Physiol 86.2:255–264

Kilarski WM, Rothery S, Roomans GM, Ulmsten U, Rezapour M, Stevenson S, Coppen SR, Dupont E, Severs NJ (2001) Multiple connexins localized to individual gap-junctional plaques in human myometrial smooth muscle. Microsc Res Tech 54:114–122

King EL, Altman C (1956) A schematic method of deriving the rate laws for enzyme catalyzed reactions. J Phys Chem 60:1375–1378

Kiputtayanant S, Lucas MJM, Wray S (2002) Effects of inhibiting the sarcoplasmic reticulum on spontaneous and oxytocin-induced contractions of human myometrium. Br J Obstet Gynaecol 109:289–296

Kitano H (2007) A robustness-based approach to systems-oriented drug design. Nat Rev Drug Discov 6:202–210

Kitching RP, Thrusfield MV, Taylor NM (2006) Use and abuse of mathematical models: an illustration from the 2001 foot and mouth disease epidemic in the United Kingdom. Rev Sci Tech 25(1):293–311

Klukovits A, Verli J, Falkay G, Gáspár R (2010) Improving the relaxing effect of terbutaline with phosphodiesterase inhibitors: studies on pregnant rat uteri in vitro. Life Sci. doi:10.1016/j.lfs.2010.10.010

Korovkina VP, Brainard AM, England SK (2006) Translocation of an endoproteolytically cleaved maxi-K channel isoform: mechanisms to induce human myometrial repolarization. J Physiol 573:329–341

Korovkina VP, Stamnes SJ, Brainard AM, England SK (2009) Nardilysin convertase regulates the function of the maxi-K channel isoform mK44 in human myometrium. Am J Physiol Cell Physiol 296:C433–C440

Koshland DE Jr, Némethy G, Filmer D (1966) Comparison of experimental binding data and theoretical models in proteins containing subunits. Biochemistry 5(1):365–368

Kotani M, Tanaka I, Ogawa Y, Suganami T, Matsumoto T, Muro S, Yamamoto Y, Sugawara A, Yoshimasa Y, Sagawa N, Narumiya S, Nakao K (2000) Multiple signal transduction pathways through two prostaglandin E receptor EP_3 subtype isoforms expressed in human uterus. J Clin Endocrinol Metab 85:4315–4322

Kryzhanovskaya-Kaplun EF, Martynshin MY (1974) Electrohysterography based on recording fast uterine potentials in women. Bull Eksper Biol Med 78:8–11 (in Russian)

Kuriyama H, Kitamura K, Itoh T, Inoue R (1998) Physiological features of visceral smooth muscle cells, with special reference to receptors and ion channels. Physiol Rev 78:811–920

La Rosa PS, Eswaran H, Preissl H, Nehorai A (2009) Forward modeling of uterine EMG and MMG contractions. Proceedings of the 11th World Congress on Medical Physics and Biomedical Engineering, Munich, Germany

La Rosa PS, Nehorai A, Eswaran H, Lowery CL, Preissl H (2008) Detection of uterine MMG contractions using a multiple change point estimator and the K-means cluster algorithm. IEEE Trans Biomed Eng 55:453–467

Ledoux J, Greenwood IA, Leblanc N (2005) Dynamics of Ca^{2+} – dependent Cl^- channel modulation by niflumic acid in rabbit coronary arterial myocytes. Mol Pharmacol 67:163–173

Lee JH, Daud AN, Cribbs LL, Lacerda AE, Pereverzev A, Klockner U, Schneider T, Perez-Reyes E (1999) Cloning and expression of a novel member of the low voltage-activated T-type calcium channel family. J Neurosci 19:1912–1921

Lee SE, Ahn DS, Lee YH (2009a) Role of T-type Ca^{2+} channels in the spontaneous phasic contractions of pregnant rat uterine smooth muscle. Korean J Physiol Pharmacol 13:241–249

Lee SL, Tan E, Khullar V, Gedroyc W, Darzi A, Yang GZ (2009b) Physical-based statistical shape modeling of the levator ani. IEEE Trans Med Imaging 28(6):926–936

Leppert PC, Yu SY (1991) Three-dimensional structures of uterine elastic fibers: scanning electron microscopic studies. Connect Tissue Res 27:15–31

Leroy MJ, Mehats C, Duc-Goiran P, Tanguy G, Robert B, Dallot E, Mignot TM, Grance G, Ferre F (1999) Effect opf pregnancy on PDE4 cAMP-specific phosphodiesterase messenger ribonucleic acid expression in human myometrium. Cell Signal 11:31–37

Li X, Kruger J, Nash M, Nielsen P (2009) Modeling fetal head motion and its mechanical interaction with the pelvic floor during childbirth. 39th Annual meeting of the international continence society, San Francis, USA. http://www.ses.auckland.ac.nz/uoa/home/about/our-research-2/publications-15/2009-publications

Li X, Kruger JA, Nash MP, Nielsen PMF (2009b) Modeling childbirth: elucidating the mechanisms of labor. WIREs Syst Biol Med 2:460–470

Li X, Kruger JA, Nash MP, Nielsen PMF (2010) effects of nonlinear muscle elasticity on pelvic floor mechanics during vaginal childbirth. J Biomech Eng 132:111010

Li Y, Gallant C, Malek S, Morgan KG (2007) Focal adhesion signaling is required for myometrial EKR activation and contractile phenotype switch before labor. J Cell Biochem 100:129–140

Li Y, Je HD, Malek S, Morgan KG (2003) ERK1/2-mediated phosphorylation of caldesmon during pregnancy and labor. Am J Physiol 284:R192–R199

Li Y, Reznichenko M, Tribe RM, Hess P, Taggart M, Kim H, DeGnore JP, Gangopadhyay S, Morgan KG (2009c) Stretch activates human myometrium via ERK, caldesmon and focal adhesion signaling. PLoS One 4(10):e7489

Liang Z, Sooranna SR, Engineer N, Tattersalt M, Khanjani S, Bennet PR, Myatt L, Johnson MR (2008) Prostaglandin F2-alpha receptor regulation in human uterine myocytes. Mol Hum Reprod 14:215–223

Lien K-C, DeLancey JOL, Ashton-Miller JA (2009) Biomechanical analyses of the efficacy of patterns of maternal effort on second stage prgroess. Obstet Gynecol 103:873–880

Lin PC, Li X, Lei ZM, Rao CV (2003) Human cervix contains functional luteinizing hormone/human chorionic gonadotropin receptors. J Clin Endocrinol Metab 88(7):3409–3414

Liu YL, Nwosu UC, Rice PJ (1998) Relaxation of isolated human myometrial muscle by β_2 -adrenergic receptors but not β_1 -adrenergic receptors. Am J Obst Gynecol 179:895–898

Longbottom ER, Luckas MJM, Kupittayanant S, Badrick E, Shmigol A, Wray S (2000) The effects of inhibiting myosin light chain kinase on contraction and calcium signaling in human and rat myometrium. Pflügers Arch 440:315–321

Loudon JAZ, Sooranna SR, Bennett PR, Johnson MR (2004) Mechanical stretch of human uterine smooth muscle cells increases IL-8 mRNA expression and peptide synthesis. Mol Hum Reprod 10:895–899

MacIntyre DA, Tyson EK, Read M, Smith R, Yeo G, Kwek K, Chan E-C (2008) Contraxction in human myometrium is associated with changes in small heart shock proteins. Endocrinology 149(1):245–252

MacPhee DJ, Lye SJ (2000) Focal adhesion signaling in the rat myometrium is abruptly terminated with the onset of labor. Endocrinology 141(1):274–283

Manoogian SJ, McNally C, Stitzel JD, Duma SM (2008) Dynamic biaxial tissue properties of pregnant porcine uterine tissue. Stapp Car Crash J 52:167–185

Marque CK, Terrien J, Rihana S, Germain G (2007) Preterm labor detection by use of a bio-physical marker: the uterine electrical activity. BMC Pregnancy Childbirth. doi:10.1186/14771-2393-7-S1-S5

Martins JAC, Pato MPM, Pires EB, Jorge RMN, Parente M, Mascarenhas T (2007) Finite element studies of the deformation of the pelvic floor. Ann NY Acad Sci 1101:316–334

Mehats C, Schmitz T, Oger S, Hervé R, Cabrol D, Leroy M-J (2007) PDE4 as a target in preterm labour. BMC Pregnancy Childbirth. doi:10.11186/1471-2393-7-S1-S12

Mehats C, Tanguy G, Paris B, Robert B, Pernin N, Ferre F, Leroy MJ (2000) Pregnancy induces a modulation of the cAMP phosphodiesterase 4-conformers ration in human myometrium: consequences for the utero-relaxant effect of PDE4-selective inhibitors. J Pharmacol Exp Ther 292:817–823

Merighi A (2002) Costorage and coexistence of neuropeptides in the mammalian CNS. Prog Neurobiol 66:161–190

Mershon JL, Mikala G, Schwartz A (1994) Changes in the expression of L-type calcium channel during pregnancy and parturition in rat. Biol Reprod 51:993–999

Metaxa-Mariatou V, McGavigan CJ, Robertson K, Stewart C, Cameron IT, Campbell S (2002) Elastin distribution in the myometrial and vascular smooth muscle in the human uterus. Mol Hum Reprod 8:559–565

Mhaouty-Kodja S, Bouet-Alard R, Limon-Boulez I, Maltier JP, Legrand C (1997) Molecular diversity of adenylyl cyclases in human and rat myometrium. J Biol Chem 272:31100–31106

Mhaouty-Kodja S, Houdeau E, Legrand C (2004) Regulation of myometrial phospholipase C system and uterine contraction by β-adrenergic receptors in midpregnant rat. Biol Reprod 70: 570–576

Miftahof R, Nam HG, Wingate DL (2009) Mathematical modeling and numerical simulation in enteric neurobiology. World Scientific Publishing, New Jersey

Miftahof RN, Nam HG (2010) Mathematical foundations and biomechanics of the digestive system. Cambridge University Press, Cambridge

Mizrahi J, Karni Z (1975) A mechanical model for uterine muscle activity during labor and delivery. Israel J Technol 13:185–191

Mizrahi J, Karni Z (1981) A constitutive equation for isotropic smooth muscle. Israel J Technol 19:143–146

Mizrahi J, Karni Z, Polishuk WZ (1978) A kinematic analysis of uterine deformation during labor. J Franklin Inst 306:119–132

Molnar M, Hertelendy F (1990) Regulation of intracellular free calcium in human myometrium. J Clin Endocrinol Metab 85:3468–3475

Monod J, Wyman J, Changeux JP (1965) On the nature of allosteric transitions: a plausible model. J Mol Biol 12:88–118

Monteil A, Chemin J, Leuranguer V, Altier C, Mennessier G, Bourinet E, Lory P, Nargeot J (2000) Specific properties of T-type calcium channels generated by the human aII subunit. J Biol Chem 275:16530–16535

Moss SB, Getton GL (2001) A-kinase anchor proteins in endocrine systems and reproduction. Trends Endocrinol Metab 12:434–440

Mosser V, Amana JIJ, Schimerlik MI (2002) Kinetic analysis of M_2 muscarinic receptor activation of G_i in Sf9 insect cell membranes. J Biol Chem 227:922–931

Nagumo J, Animoto S, Yoshigawa S (1962) An active pulse transmission line simulating nerve axon. Proc Inst Radio Eng 50:2061–2070

Nakanishi H, Wood C (1971) Cholinergic mechanisms in the human uterus. J Obst Gynaecol Br Commonw 78(8):716–723

Noakes KF, Pullan AJ, Bissett IP, Cheng LK (2008) Subject specific finite elasticity simulations of the pelvic floor. J Biomech 41:3060–3065

Noble D (2002) Modeling the heart – from genes to cells to the whole organ. Science 295: 1678–1682

Norden AP (1950) In relation to the theory of finite deformations. Proc Kazan Branch Acad Sci USSR Ser Phys Maths Tech Sci 2:12–37

Nusbaum MP, Blitz DM, Swensen AM, Wood D, Marder E (2001) The roles of co-transmission in neural network modulation. Trends Neurosci 24(3):146–154

Oger S, Mehats C, Barnette MS, Ferre F, Cabrol D, Leroy M-J (2004) Biol Reprod 70:458–464

Oldenhof AD, Shynlova OP, Liu M, Langille BL, Lye SJ (2002) Mitogen-activated protein kinases mediate stretch-induced c-fos mRNA expression in myometrial smooth muscle cell. Am J Physiol Cell Physiol 283:C1530–C1539

Olson DM, Zaragoza DB, Shallow MC, Cook JL, Mitchell BF, Grigsby P, Hirst J (2003) Myometrial activation and preterm labour: evidence supporting a role for the prostaglandin F receptor – a review. Placenta 24:S47–S54

Omhichi M, Koike K, Nohara A, Kanda Y, Sakamoto Y, Zhang ZX, Hirota K, Miyake A (1995) Oxytocin stimulates mitogen-activated protein kinase activity in cultured human puerperal uterine myometrial cells. Endocrinology 136:2082–2087

Osmers RG, Adelmann-Grill BC, Rath W, Stuhlatz HW, Tchesche H, Kuhn W (1995a) Biocehmical events in cervical ripening dilatation during pregnancy and parturition. Am J Obstet Gynecol 166:1455–1460

Osmers RG, Balser J, Kuhn W, Tchesche H (1995b) Interleukin-8 synthesis and the onset of labor. Obstet Gynecol 86:223–229

Parco IE-L, Dallot E, Breuiller-Fouché M (2007) Protein kinase C and human uterine contractility. BMC Preg Childbirth. doi:10.1186/1471-2393-7-S1-S11

Parente MP, Jorge N, Renato M, Mascarenhas T, Silva-Filho Agnaldo L (2010) The influence of pelvic muscle activation during vaginal delivery. Obstet Gynecol 115:804–808

Parkington HC, Coleman HA (2001) Excitability in uterine smooth muscle. Front Horm Res 27:179–200

Parkington HC, Tonta MA, Brennecke S, Coleman HA (1999) Contractile activity, membrane potential, and cytoplasmic calcium in human uterine smooth muscle in the third trimester of pregnancy and during labor. Am J Obstet Gynecol 181:1445–1451

Paskaleva AP (2007) Biomechanics of cervical function in pregnancy – case of cervical insufficiency. PhD Thesis Dept Mech Eng Mass Inst Tech, USA 212 p

Patak E, Candenas ML, Pennefather JN, Ziccone S, Lilley A, Martin JD, Flores C, Mantecon AG, Story ME, Pinto FM (2003) Tachykinins and tachykinin receptors in human uterus. Br J Pharmacol 139:523–532

Patak E, Ziccone S, Story ME, Fleming AJ, Lilley A, Pennefather JN (2000) Activation of neurokinin NK2 receptors by tachykininn peptides casues contraction of uterus in pregnant women near term. Mol Hum Reprod 6:549–554

Pearsall GW, Roberts VL (1978) Passive mechanical properties of uterine muscle (myometrium) tested in vitro. J Biomech 11:167–176

Pieber D, Allport VC, Hills F, Johnson M, Bennett PR (2001) Interactions between progesterone receptor isoforms in myometrial cells in **hum**an labour. Mol Hum Reprod 7:875–879

Planes JG, Morucci JP, Grandjean H, Favretto A (1984) External recording and processing of fast electrical activity of the uterus in human parturition. Med Biol Eng Comput 22(6):585–591

Plonsey RL, Barr RG (1984) Current flow patterns in two-dimensional anisotropic bisyncytia with normal and extreme conductivities. Biophys J 43:557–571

Popescu LM, Ciontea SA, Cretoiu D (2007) Interstitial Cajal-like cells in human uterus and fallopian tubes. Ann NY Acad Sci 1101:139–165

Premont RT, Chen J, Ma HW, Ponnapalli M, Iyengar R (1992) Two members of a widely expressed subfamily of hormone-stimulated adenylyl cyclases. Proc Natl Acad Sci USA 89:9809–9813

Pressman EK, Tucker JA, Anderson NC Jr, Young RC (1988) Morphologic and electrophysiologic characterization of isolated pregnant human myometrial cells. Am J Obstet Gynecol 159(5): 1273–1279

Price SA, Bernal AL (2001) Uterine quiescence: the role of cyclic AMP. Exp Physiol 86(2):265–272

Prud'Home M-J (1986) Uterine motor responses to an α-adrenergic agonist (phenylephrine) in the ewe during oestrus and at the end of gestation. Reprod Nutr Dev 26(3):827–839

Ramon C, Preissl H, Murphy P, Wilson JD, Lowery C, Eswaran H (2005a) Synchronization analysis of the uterine magnetic activity during contractions. Biomed Eng. doi:10.1186/1475-925X-4-55

Ramon C, Preissl H, Murphy P, Wilson JD, Lowery C, Hari E (2005b) Synchronization analysis of the uterine magnetic activity during contractions. Biomed Eng. doi:10.1186/1475-925X-4-55, 1-12

Ramsey ME (1994) Anatomy of the human uterus. In: Chard T, Grudzinskkas G (eds) The uterus. Cambridge University Press, Cambridge, pp 18–29

Rasmussen BB, Larsen LS, Senderovitz T (2005) Pharmacokinetic interaction studies of atosiban with labetalol or betamethasone in healthy female volunteers. Int J Obstet Gynecol. doi:10.1111/ j.1471-0528.2005.00735.x

Rauwerda H, Roos M, Herttzberger BO, Breit TM (2006) The promise of a virtual lab. DDT 11: 228–236

Reinheimer TM (2007) Barusiban suppresses oxytocin-induced preterm labour in non-human primates. BMC Pregnancy Childbirth. doi:10.1186/1471-2393-7-S1-S15

Reinheimer TM, Bee WH, Resendez JC, Meyer JK, Haluska GJ, Chellman GJ (2005) Barusiban, a new highly potent and long acting oxytocin antagonist: pharmacokinetic and pharma-codynamic comparison with atosiban in a cynomolgus monkey model of preterm labor. J Clin Endocrinol Metab 90:2275–2281

Reynolds SRM (1949) Gestational mechanisms. In: Physiology of the uterus, 2nd edn. Hoeber, New York, pp 218–234

Rezapour M, Backstrom T, Ulmsten U (1996) Myometrial steroid concentration and oxytocin receptor density in parturient women at term. Steroids 61(6):338–344

Rice D, Yang T (1976) A nonlinear viscoelastic membrane model applied to the human cervix. J Biomech 9:201–210

Rice D, Yang T, Stanley P (1975) A simple model of the human cervix during the first stage of labor. J Biomech 9:153–163

Richter ON, Dorn C, van de Vondel P, Ulrich U, Schmolling J (2005) Tocolysis with atosiban: experience in the manamgement of premature labor before 24 weeks of pregnancy. Arch Gynecol Obstet 272:26–30

Ridel VV, Gulin BV (1990) Dynamics of soft shells. Nauka, Moscow

Rihana S, Lefrançois E, Marque C (2007) A two dimensional model of the uterine electricalwave propagation. In: Proceedings of the 29th annual international conference of the IEEE EMBS, Lyon, France, August 23–26, pp 1109–1112

Rihana S, Santos J, Marque C (2006) Dynamical analysis of uterine cell electrical activity model. In: Proceedings of the 28th annual international conference of the IEEE EMBS, New York, USA, August 30–September 3, pp 4179–4182

Rihana S, Terrien J, Germain G, Marque C (2009) Mathematical modeling of electrical activity of uterine muscle cells. Med Biol Eng Comput 47:665–675

Robinson EE, Foty RA, Corbett SA (2004) Fibronectin matrix assembly regulates alpha5 beta1-mediated cell adhesion. Mol Biol Cell 15:973–981

Roh CR, Lee BL, Oh WJ, Whang JD, Choi DS, Yoon BK, Lee JH (1999) Introduction of c-Jun mRNA without changes of estrogen and progesterone receptor expression in myometrium during human labor. J Korean Med Sci 14:552–558

Romero R, Sibai BM, Sanchez-Ramos L, Valenzuela GJ, Veille JC, Tabor B, Perry KG, Varner M, Goodwin TM, Lane R (2000) An oxytocin receptor antagonist (atosiban) in the treatment of preterm labor: randomized, double-blind, placebo-controlled trial with tocolytic rescue. Am J Obstet Gynecol 182:1173–1183

Rouget C, Bardou M, Breuiller-Fouché M, Loustalot C, Qi H, Naline E, Croci T, Carbol D, Advenier C, Leroy MJ (2005) β_3-Adrenoceptor is the predominant β-adrenoceptors type in human myometrium and its expression is regulated in pregnancy. J Clin Endocrinol Metab 90:1644–1650

Sakai N, Tabb T, Garfiled R (1992) Studies of connexin 43 and cell-to-cell coupling between myometrial cells of the human uterus during pregnancy. Am J Obstet Gynecol 167:1267–1277

Salomonis N, Cotte N, Zambon AC, Pollard KS, Vranizan K, Doniger SW, Dolganov G, Conklin BR (2005) Identifying gene networks underlying myometrial transition to labor. Genome Biol 6:R12

Sanborn B (2007) Hormonal signaling and signal pathway crosstalk in the control of myometrial calcium dynamics. Semin Cell Dev Biol 18:305–314

Sanborn BM, Dodge KL, Monga M, Quian A, Wang W, Yue C (1998a) Molecular mechanisms regulating the effects of oxytocin on myometrial intracellular calcium. Adv Exp Med Biol 449:277–286

Sanborn BM, Ku C-Y, Shlykov S, Babicj L (2005) Molecular signaling through G-protein-coupled receptors and the control of intracellular calcium in myometrium. J Soc Gynecol Investig 12:479–487

Sanborn BM, Yue C, Wang W, Dodge KL (1998b) G protein signaling pathways in myometrium: affecting the balance between contraction and relaxation. Rev Reprod 3:196–205

Seitchik J, Chatkoff ML (1975) Intrauterine pressure wave-form characteristics of spontaneous first stage labor. J Appl Physiol 38:443–448

Serradeil-Le Gal C, Valette G, Foulon L, Germain G, Advenier C, Naline E, Bardou M, Martinolle JP, Pouzet B, Raufaste D, Garcia C, Double-Cazanave E, Pauly M, Pascal M, Barbier A, Scatton B, Maffrand JP, Le Fur G (2004) SSR126768A (4-chloro-3-[(3R)-(+)-5-chloro-1-(2,4-dimethoxybenzyl)-3-methyl-2-oxo-2,3-dihydro-1H-indol-3-yl]-N-ethyl-N-(3-pyridylmethyl)-benzamide, hydrochloride): a new selective and orally active oxytocin receptor antagonist for the prevention of preterm labor. J Pharmacol Exp Ther 309(1):414–24

Sfakiani A, Buhimschi I, Pettker C, Magliore L, Turan O, Hamer B, Buhimschi C (2008) Ultrasonographic evaluation of myometrial thickness in twin pregnancies. Am J Obstet Gynecol 198(5):530, e1–10

Shlykov SG, Sanborn BM (2004) Stimulation of intracellular Ca^{2+} oscillations by diacyglycerol in human myometrial cells. Cell Calcium 36:157–164

Shynlova O, Williams SJ, Draper H, White BG, MacPhee DJ, Lye SJ (2007) Uterine stretch regulates temporal and spatial expression of fibronectin protein and its alpha 5 integrin receptor in myometrium of unilaterally pregnant rats. Biol Reprod 77:880–888

Slattery MM, Morrison JJ (2002) Preterm delivery. Lancet 360:1489–1497

Small JV, Gimona M (1998) The cytoskeleton of the vertebrate smooth muscle cell. Acta Physiol Scand 164:341–348

Smith PG, Garcia R, Kogerman L (1997) Strain reorganizes focal adhesion and cytoskeleton in cultured airway smooth muscle cells. Exp Cell Res 232:127–136

Sooranna SR, Grigsby P, Myatt L, Bennett PR, Johnson MR (2005) Prostanoid receptors in human uterine myocytes: the effect of reproductive state and stretch. Mol Hum Reprod 11:859–864

Sooranna SR, Lee Y, Ki LU, Mohan AR, Bennett PR, Johnson MR (2004) Mechanical stretch activates type 2 cyclooxigenase via activator protein-1 transcription factor in human myometrial cells. Mol Hum Reprod 10:109–113

Stockton JM, Birdstall NJM, Burgen ASV, Hulme EC (1983) Modification of the binding-properties of muscarinic receptors by gallamine. Mol Pharmacol 23:551–557

Strachan RT, Ferrara G, Roth BL (2006) Screening the receptorome: an efficient approach for drug discovery and target validation. Drug Discov Today 11:708–716

Stratova Z, Soloff MS (1997) Coupling of oxytocin receptor to G proteins in rat myometrium during labor: Gi interaction. Am J Physiol 272:E870–E876

Tabb T, Thilander G, Grover A, Hertzberg E, Garfield R (1992) An immunochemical and immunocytologic study of the increase in myometrial gap junctions (connexin 43) in rats and humans during pregnancy. Am J Obstet Gynecol 167:559–567

Taggart MJ, Wray S (1998) Contribution of sarcoplasmic reticular calcium to smooth muscle contractile activation: gestational dependence in isolated rat uterus. J Physiol 511:133–144

Teschemacher AG, Christopher DJ (2008) Cotransmission in the autonomic nervous system. Exp Physiol 94(1):18–19

Thomas R (2009) Kidney modeling and systems physiology. WIREs Syst Biol Med 1:172–190. doi:10.1002/wsbm.014

Tribe RM (2001) Regulation of human myometrial contractility during pregnancy and labor: are calcium homeostatic pathways important? Exp Physiol 86(2):247–254

Trudeau L-E, Gutiérrez R (2007) On cotransmission & neurotransmitter phenotype plasticity. Mol Interv 7:138–146

Tucek S, Musilkova J, Nedoma J, Proska J, Shelkovnikov S, Vorlicek J (1990) Positive cooperativity in the binding of alcuronium and N-methylscopamine to muscarinic acetylcholine receptors. Mol Pharmacol 38:674–680

Usik PI (1973) Continual mechanochemical model of muscular tissue. Appl Math Mech 37:448–458

Vauge C, Carbonne B, Papiernik E, Ferré F (2000) A mathematical model of uterine dynamics and its application to human parturition. Acta Biotheor 48:95–105

Vauge C, Mignot T-M, Paris B, Breulier-Fouché M, Chapron C, Attoui M, Ferré F (2003) A mathematical model for the spontaneous contractions of the isolated uterine smooth muscle from patients receiving progestin treatment. Acta Biotheor 51:19–34

Weiss JA, Gardiner JC (2001) Computational modeling of ligament mechanics. Crit Rev Biomed Eng 29:303–371

Weiss S, Jaermann T, Scmid P, Staempfli P, Boesiger P, Niederer P, Caduff R (2006) Three dimensional fiber architecture of the non-pregnant human uterus determined ex vivo using magnetic resonance diffusion tensor imaging. Anat Rec A Discov Mol Cell Evol Bio 288:84–90

Weiss ST, Bajka M, Nava A, Mazza E, Niederer P (2004) A finite element model for the simulation of hydrometra. Technol Health Care 12:259–267

Wex J, Connolly M, Rath W (2009) Atosiban versus betamimetics in the treatment of preterm labour in Germany: an economic evaluation. BMC Pregnancy Childbirth. doi:10.1186/1471-2393-9-23

Willets JM, Taylor AH, Shaw H, Konje JC, Challiss RAJ (2008) Selective regulation of H1 histamine receptor signaling by G protein-coupled receptor kinase 2 in uterine smooth muscle cells. Mol Endocrinol 22(8):1893–1907

Williams SJ, White B, MacPhee DJ (2005) Expression of $\alpha 5$ integrin (Itga5) is elevated in the rat myometrium during late pregnancy and labor: implications for development of a mechanical syncytium. Biol Reprod 72:1114–1124

Wolfs GMLA, Rottinghuis H (1970) Electrical and mechanical activity of the human uterus during labour. Arch Gynäkol 208:373–385

Wood C (1964a) Physiology of uterine contractions. J Obstet Gynaecol Br Commonw 71:360–373

Wood C (1964b) The expansile behavior of the human uterus. J Obstet Gynaecol Br Commonw 71:615–620

Wood JR, Likhite VS, Loven MA, Nardulli AM (2001) Allosteric modulation of estrogen receptor conformation by different estrogen response elements. Mol Endocrinol 15:1114–1126

Woodward TL, Mienaltowski AS, Bennett JM, Haslam SZ (2001) Fibronectin and the alpha(5) beta(1) integrin are under developmental and ovarian steroid regulation in the normal mouse mammary gland. Endocrinology 142:3214–3222

Wu X, Morgan KG, Jones CJ, Tribe RM, Taggart MJ (2008) Myometrial mechanoadaptation during pregnancy: implications for smooth muscle plasticity and remodeling. J Cell Mol Med 12(4):1360–1373

Yamagata M, Sanes JR, Weiner JA (2003) Synaptic adhesion molecules. Curr Opin Cell Biol 15:621–632

Yamamoto T, Miyamoto S (1995) Integrin transmembrane signaling and cytoskeletal control. Curr Opin Cell Biol 7:681–689

Yoshimura M, Cooper DM (1992) Cloning and expression of a Ca(2+)-inhibitable adenylyl cyclase from NCB-20 cells. Proc Natl Acad Sci USA 89:6716–6720

Young R, Hession R (1999) Three-dimensional structure of the smooth muscle in the term-pregnant human uterus. Obstet Gynecol 93:94–99

Young RC (2007) Myocytes, myometrium, and uterine contractions. Ann NY Acad Sci 1101:72–84

Young RC, Henrdon-Smith L (1991) Characterization of sodium channels in cultured human uterine smooth muscle cells. Am J Obstet Gynecol 164(1):175–181

Young RC, Smith LH, MacLaren MD (1993) T-type and L-type calcium currents in freshly dispersed human uterine smooth muscle cells. Am J Obstet Gynecol 195:1404–1406

Young RC, Zhang P (2005) Inhibition of in vitro contractions of human myometrium by mibefradil, a T-type calcium channel blocker: support for a model using excitation-contraction coupling, and autocrine and paracrine signaling mechanisms. J Soc Gynecol Investig 12:e7–e12

Yu JT, Bernal L (1998) The cytoskeleton of human myometrial cells. J Reprod Fertil 112:185–198

Yuan W, Bernal AL (2007) Cyclic AMP signaling pathways in the regulation of uterine relaxation. BMC Pregnancy Childbirth. doi:10.1186/1471-2393-7-S1-S10

Yue C, Dodge KL, Weber G, Sanborn BM (1998) Phosphorylation of serine 1105 by protein kinase A inhibits Cbeta3 stimulation by Galpaq. J Biol Chem 273:18023–18027

Zhou XL, Lei ZM, Rao CV (1999) Treatment of endometrial gland epithelial cells with chorionic gonadotrophin/luteinizing hormone increases the expression of the cyclooxygenase-2 gene. J Clin Endocrinol Metab 84:3364–3377

Ziganshin A, Zefirova JT, Zefirova TP, Ziganshina LE, Hoyle CHV, Burnstock G (2005) Potentiation of uterine effects of prostaglandin $F_{2\alpha}$ by adenosine 5′-triphosphate. Obstet Gynecol 105(6):1429–1436

Ziganshin AU, Zaitcev AP, Khasanov AA, Shamsutdinov AF, Burnstock G (2006) Term-dependency of P2 receptor-mediated contractile responses of isolated human pregnant uterus. Eur J Obstet Gynecol Reprod Biol 129(2):128–134

Index